STOCHASTIC THEORY

OF

SERVICE SYSTEMS

by

L. KOSTEN
Delft Institute of Technology,
The Netherlands

PERGAMON PRESS

OXFORD · NEW YORK · TORONTO

SYDNEY · BRAUNSCHWEIG

Pergamon Press Ltd., Headington Hill Hall, Oxford
Pergamon Press Inc., Maxwell House, Fairview Park, Elmsford, New York 10523
Pergamon of Canada Ltd., 207 Queen's Quay West, Toronto 1
Pergamon Press (Aust.) Pty. Ltd., 19a Boundary Street, Rushcutters Bay, N.S.W. 2011, Australia
Vieweg & Sohn GmbH, Burgplatz 1, Braunschweig

First edition 1973

Library of Congress Cataloging in Publication Data

Kosten, Leendert.
Stochastic theory of service systems.

(International series of monographs in pure and applied mathematics)
Bibliography: p.
1. Queuing theory. 2. Stochastic processes.
I. Title.
T57.9.K67 658.4'032 72-10124
ISBN 0-08-016948-1

Printed in Germany

INTERNATIONAL SERIES OF MONOGRAPHS IN

PURE AND APPLIED MATHEMATICS

General Editors: I. N. Sneddon and M. Stark

Executive Editors: J. P. Kahane, A. P. Robertson and S. Ulam

Volume 103

STOCHASTIC THEORY OF SERVICE SYSTEMS

TO MARGOT

CONTENTS

PREFACE

THE stochastic theory of service systems came into existence several decades ago in the area of telephony (Erlang, about 1910). Telephone service systems comprise groups of "servers" who offer their services—usually on a one-at-a-time basis—to groups of customers, the "(traffic) sources". The groups of sources are usually much larger than the group of servers. The sources are non-co-operative, and their demands for service are not necessarily distributed uniformly over time. The demands may form clusters in a haphazard way. It then is possible that the number of servers may not be large enough to handle the demands in those periods of "congestion". It is one of the objects of the theory under consideration to establish and calculate measures for this congestion.

For some decades the theory was developed almost exclusively in the area of telecommunications. In this area the models are more or less self-evident and robust, with well-defined technical aggregates of servers functioning in a completely deterministic way, whilst the behaviour of the sources (the subscribers) is not deterministic but is very well known in a statistical sense. The development of the theory in telephony for some time showed rather isolationist tendencies, the early start being one cause, but there are other factors. Publications mostly appeared in the home periodicals of administrative units and of a small number of large industries. Although those periodicals were of a high quality they were not widely known outside the telecommunications world. International organizations in this field exhibit to some extent a bit of a "closed-shop" structure, which since 1955 has led to the so-called "International Teletraffic Congresses" (ITC): 1955 Copenhagen; 1958 The Hague; 1961 Paris; 1964 London; 1967 New York; 1970 Munich; (1973, seventh ITC, Stockholm).

There have been a few laudable efforts to break this relative isolation. In 1948 the Copenhagen Telephone Administration (KTAS) took the initiative of publishing Erlang's complete scientific works (Brockmeyer *et al.*, 1948). The Automatic Telephone and Electric Co. Ltd. generously sponsored the publication of Syski's *Congestion Theory in Telephone Systems* (1960), one of the standard works.

Since about 1950 there has been a marked development of the use of Congestion Theory in other fields: all kinds of physical traffic, workshops, inventories, medical care, data handling, etc., have been discussed. Automation and its consequent integration yielded larger and less vague systems in those fields, the functioning of which was better understood. Moreover, the increasing size of those systems made it impossible to apply rigorous systemwide planning and scheduling. This led to the concept of analysing a system (an industrial one, say) into a number of subsystems, which internally may function deterministically, but the interaction of which entails stochastic congestion problems.

Though the problems at hand in telecommunication and in other fields are, roughly speaking, alike, there are a few striking differences. In telecommunications the "holding-times" (durations of conversations) are not known in advance. In various other areas, however, the period of occupation of the server is often connected with some job to be done, the duration of which may pretty well be known in advance. This causes a marked interest in "priority rules" in such areas, an interest that is not to be found in telecommunications.

In telecommunications (including data handling) the servers are (usually) strictly technical structures. In other areas, however, the servers themselves are "man–machine" aggregates. This means that in those cases the model must include assumptions as to the behaviour of man in those aggregates, e.g. "according to given rules" or "rationally" (whatever this may mean). Those assumptions are necessarily questionable. Consequently, the results of investigations are more speculative here than in the field of telecommunication. Hence, applications in non-telecommunication areas tend to deal with small systems, having one or a few servers only. In telecommunications, however, elaborate systems with many servers, interacting in a very subtle way, are often investigated.

In accordance with the picture given above the problem field of congestion theory consists of a few standard cases (normally covered by the magic word "Erlang"), surrounded by a great variety of modifications that are important in special areas of application. The present book deals with the standard cases as well as with some problem variants that are not too specialized to be understood by the general reader, to whom they may be of interest. The author's background in telephony may partially account for the choice of subject-matter.

There exists a unifying theory in the Theory of Stochastic Processes. It is a rather abstract body of knowledge, indispensable for those who are interested in questions about existence and uniqueness of solutions. It is, however, beyond the scope of this book. We shall stick to the engineer's viewpoint, that apparently sound problems as a rule have a (unique) solution and that nature will warn us by anomalies in the calculations if we are wrong in this assumption.

It is an aim of this book to show that relatively simple analytical means are sufficient in many cases. As far as possible the method of straightforward formulation and solution of "birth-and-death equations" (or "equations of state") will be employed. In fact those equations are identical with the forward Chapman–Kolmogorov equations, familiar in the theory of Markov processes. Their use, however, goes back to Erlang (1918, 1925) and Molina (1927). In more complicated cases it is necessary to "markovize" the processes by the introduction of supplementary variables, in order to be able to formulate those equations.

In a few of the later chapters the practical limitations of the method of equations of state will become apparent. It is shown that the use of other—combinatorial—methods may sometimes be more fruitful.

Analysis may in many cases lead to very complex results. In pre-computer times there was not the slightest possibility of obtaining numerical results in any but the simplest cases. It is not surprising that elegance of analytical results counted in those days. Nowadays, however, it may be asked that analysis be conducted in such a direction that numerical results may be obtained by computer in as easy a way as possible. This means that no longer should we aim at obtaining closed expressions but rather at deriving practicable algorithms. In many instances this leads to direct numerical integration of the equations of state.

Often in practice the situation is so complex as to preclude analysis or even numerical methods. This does not mean that theory is then useless. In many cases it proves possible to split up large systems into subsystems, the behaviour of which can be dealt with in an approximate way by known theoretical results. Besides a sound knowledge of theory this requires considerable skill that can only be acquired in practice.

In complicated cases there fortunately is a method of ultimate resource, viz. simulation. A chapter of this book has been devoted to simulation for two reasons. In the first place a special type of simulation, "roulette simulation", is used in telecommunications, and this needs some theoretical background, to be found in congestion theory. Secondly, a discussion of the accuracy of results, obtained by simulation, gives rise to a type of analysis that is very similar to that of congestion theory.

Prerequisites for this book are elementary statistics and probability theory, an engineer's course in calculus besides some notion of a function of a complex variable and of Laplace Transforms. The use of generating functions is explained in an appendix. Most of the book—notably the central chapters—can be understood on the basis of rather elementary mathematics.

I wish to thank Mr. T. C. A. Mensch for valuable criticism and accurate proof-reading; Miss C. C. M. Hoogervorst and Miss I. V.van Hoogdalem for the careful typing of the manuscript; Mr. S. J. Ahern and the editor's referee (formally unkown to me) for making the text look real English; the latter also for a vauable suggestion concerning the matter of the book; and finally the editorial staff of Pergamon Press for a good co-operation.

L. K.

1. INTRODUCTION

1.1. General description of service systems

In the beginning of this century telephone engineers badly needed rules for determining the numbers of connecting lines, other equipment, operators, etc., necessary to handle the ever-increasing traffic. It was required that the traffic should be handled adequately, but in as economical a way as possible. Up to that time rules of thumb had been used. The first theory that had a mathematical foundation (viz. the theory of A. K. Erlang, 1918, 1925) immediately proved to be a success. Erlang's formulae, and more generally his approach, are at the basis of many current contributions.

The prototype of the problems encountered in this field is the following. There are a number c of identical telephone lines leading from a town A to a town B. Each time a subscriber in A desires a telephone connection with a subscriber in B, a path is set up comprising, *inter alia*, one of the c lines mentioned. The line is occupied as long as the connection lasts and is released immediately afterwards. Now, the number of subscribers in A is usually much larger than c. Hence, under unfavourable conditions it may happen that the number of simultaneously desired connections from A to B exceeds the number c of available lines. This means inconvenience to the subscribers in A. Depending on the way the system has been engineered, this hindrance has one of two possible aspects. A demand made when all c lines are engaged will either:

(i) never be given service; the dialling subscriber is informed of this fact by an "*engaged*" *signal* or *busy-tone* (*busy-tone systems* or *blocking systems*); the demands that are not given service are said to encounter *blocking*; or

(ii) it will be given service as soon as a line becomes available again; these *delayed demands* form a *queue* (*delay* or *queuing systems*).

Now there are many situations where transcriptions of this model apply. The calling subscribers may be replaced, for example, by car drivers trying to park their cars in a parking lot that can accommodate c cars. Or the demands for service may stem from ships nearing a port and which need one of c tug-boats to be brought in. Again, there may be c first-aid squadrons which have to deal with casualties in a certain area, and so on.

At this point it will be useful to describe briefly the meaning of some words that are used frequently. A *system* is an aggregate of entities that interact in some way. A *model* is a description of such a system that leaves out inessentials and in which the rules of interaction are stated (note that "inessential" is only meaningful when we know what aspects of the system are of interest). A *process* is a set of variables that specify the changes in time of the situation in the system or the model, so that this set of variables characterize the *behaviour* of the system or the model. A *stochastic variable* is a time-dependent variable that does not assume a fixed value at a certain time but rather one of a set of different values, according to some element of chance. When a process is non-deterministic, one or more of its characterizing variables are stochastic variables. In this case it will be called a *stochastic process*. When at every time the future values of the characterizing stochastic variables are statistically determined by their present values, the process is called a (*simple*) *Markov process*.

All the above-mentioned cases may be described by the following model of a service system. There is a *group of sources*, producing a *flow of demands* for service. The sources are not supposed to be part of the system. The flow of demands is an *input* of the system. Demands are *temporary entities* in the system. Occurrences of new demands will be called *arrivals*. The flow of demands is the *arrival process*. The service is rendered by a number (c) of identical *servers*. These servers may be in one of two states: *idle* or *engaged* (*busy*). At the arrival of a demand it is given service by an idle server, if there is one. This server changes to the engaged-state for a certain lapse of time, the *holding-time*. After this holding-time the server becomes idle again. During this holding-time the server is said to handle an *occupation*. When the demand does not find an idle server, it may take a free waiting-position in a *queue*, which can accommodate q waiting demands at a time. If all q waiting-positions in this queue are also occupied when the new demand arrives, the latter is simply discarded without further consequences. If $q = 0$ the system described is a blocking system. When $q > 0$ (delay possible) it is mostly assumed that $q = \infty$ (infinite queue). When an engaged server becomes idle at an instant at which there are waiting demands in the queue, one of those demands leaves its waiting-position (rendering the latter free) and occupies the server (which becomes engaged).

The group of sources may be a distinct set of items, as in the telephone system example. It may also be a rather vague collection: "all ships", "all cars" (in town? in the country?). The main characteristic of the group of sources is *lack of co-operation*: the sources are items that to a very large extent make their demands for service independently of each other. The consequence is that the flow of demands is rather chaotic in nature. The chaos in the system of servers is even greater, owing to the fact that the holding-times vary in a way that is not known. The burden of coping with this chaos remains the task of the exploiter of the servers.

When the sources and the servers belong to the same economic system (a factory, a firm, a community) there has always been a strong tendency towards decreasing the chaos mentioned by *planning* and *scheduling*. The sources supply a planning department with notices about anticipated demands and anticipated holding-times. By some scheduling procedure those holding-times are made to fit into some scheme that uses these servers as efficiently as possible without introducing unacceptably protracted waiting-times for the demands. In this way the procedure becomes deterministic. This system seems to be just the opposite of the aforementioned service system: total regulation versus total chaos. In practice, however, in the regulated system the actual demands and holding-times will deviate from their forecasts, again causing randomness in the system. The growing complexity of our society increases the effects of those deviations more and more, even to such an extent that the advantage of scheduling is diminished or even becomes illusory. Under those circumstances the modern philosophy in *job shops* (as distinct from standardized mass fabrication) is to consider a factory as consisting of groups of servers, e.g. groups of lathes, of drilling machines, etc. The jobs find their ways through the factory, going from one group of servers to another, until they finally leave the factory. No attempts are made to give advance notice about the routes of those jobs. When a job goes to a group of servers, it may be thought of as making a (random) demand for service. The result of dropping the scheduling is that a job may incur delays when going from one group of servers to another. Another result is that the servers occasionally have to wait for new jobs; so the *occupancy* or *efficiency* of the servers is less than 100%. It is the task of management to balance the cost of idle servers against that of delay of jobs. As a result of the concept adopted above the groups of servers are more or less decoupled: the queues of jobs waiting before the groups of servers form a kind of *buffers*. The aforementioned service-model now seems to apply for each group of servers separately. The fact that the model in question may be used for describing this modern way of organizing job-shops increases the importance of the model.

The crude model of service systems given so far needs further specification in various aspects, to be discussed in the following pages.

1.1.1. CHARACTERISTICS OF HOLDING-TIMES

The holding-time is not known beforehand; it is a so-called *stochastic variable*. It may be characterized by its distribution function, as will be explained in Section 1.2. Its mean value will be denoted by h, the inverse by $\mu = 1/h$. The quantity μ represents the average number of demands that can be given service per unit of time by one server, assuming the latter to be continuously occupied. It is therefore a measure of the *capacity* of a server: it is the so-called *service-rate*.

1.1.2. CHARACTERISTICS OF THE FLOW OF DEMANDS

This flow forms a so-called *time-series* or *point-process*, i.e. a series of instants at which demands arrive. The stochastic aspects of this time-series will be dealt with in Section 1.3. In some broad sense there is a certain expected (or average) number of arrivals per unit of time, to be denoted by λ, and termed the *arrival-rate* or *arrival-density*.

We shall frequently use a *scaling factor* h for times, i.e. h is chosen as a new unit of time. When some lapse of time is t, in the scaled version it will be denoted by $\tau = t/h$. The quantities λ and μ, having a dimension $[T^{-1}]$, vary inversely with h. Thus μ becomes $\mu h = 1$, whilst λ becomes $\lambda h = \lambda/\mu$, to be denoted by ϱ. This quantity—the scaled arrival rate or traffic density—represents the expected number of arrivals per average holding-time. The parameter ϱ is dimensionless: when the traffic density is expressed in these non-dimensional units it is said to be measured in *erlangs*.

1.1.3. THE QUEUE-DISCIPLINE

In the theory of service systems the word "queue" is used in a way that differs from common practice. When we speak of a queue in every-day life, it is tacitly assumed that the items in the queue are served in order of arrival. The concept "queue" used in the following, however, is *not* assumed to obey this assumption: the order in which waiting demands are served is purposely left open here. The set of rules concerning the queue are said to constitute the *queue-discipline*. It is described by specification of the following points:

(i) *The order in which waiting demands are given service.* This order may be governed by several aspects of the demands:

(a) *Dependent on time of origination.* The most natural prescription seems to be: "first-come-first-served", also called "first-in-first-out", which holds for most actual queuing in a physical sense (e.g. persons standing between barriers). Sometimes, however, "last-come-first-served" also called "last-in-first-out" conditions apply, e.g. physical waiting in narrow cul-de-sacs, and the piling up of documents in a drawer. Again, in telecommunication the waiting demands are frequently given service in random order, i.e. independent of time of origination. Any rule other than "random" would necessitate registration of the times of origination of the demands, which may be thought to be too expensive.

(b) *According to holding-times.* In many service systems the holding-times are not known in advance (e.g. durations of telephone calls; in gas stations it is not known beforehand whether, apart from petrol supply, it will be necessary to check oil, water and tyre-pressure). In industrial applications, however, they may sometimes be fairly well known in advance. Average waiting-times may be cut down considerably by using a rule such as SPT ("shortest processing time first").

(c) *According to origin; priority.* Sometimes the sources are not all identical, but may be distinguished in a number of different sub-groups. Subgroups produce demands with different priorities. When demands with different priorities are present in the queue, demands with a high priority will be given service before demands with a lower priority. Strictly speaking, case (b) is a form of priority. The priority is normally *non-pre-emptive.* This means that arriving high-priority demands are unable to interrupt the service that is being given to lower-priority demands. With *pre-emptive* priority, however, a high-priority demand which finds all servers busy is entitled to interrupt the service being given to a lower-priority demand. The server first helps the higher-order demand and continues or recommences the servicing of the lower-order demand afterwards.

(ii) *The possibility of leaving the queue* (balking). Normally it is assumed that waiting demands cannot leave the queue. Sometimes, however, it is assumed that impatient waiting demands can leave it. Some stochastical law, specifying the reaction on long delays, should then be given.

1.1.4. THE TIME-DEPENDENCE OF THE SYSTEM'S BEHAVIOUR; STEADY-STATE CONDITIONS

Mostly, the rate of arrival depends upon the time: the average number of customers per hour in a gas station, for example, is much smaller at night than during the day. The servers must, of course, be capable of handling the demands in the so-called *rush hour* also. During that peak-period, the density of the flow may be assumed to be more or less constant. It is therefore important to study constant arrival rate cases, with which most theoretical results are concerned. Moreover, it is usually assumed that the so-called *steady-state condition* (or *stationarity*) applies, a stronger assumption than that of constant density. It means that the probability of finding the system in a certain state is independent of the time at which the system is considered, that is that "transients" have faded away. For stationarity to apply, the arrival-rate must have been constant for a "sufficiently long" time. But what is sufficiently long? There are cases where no steady-state is possible. Let λ be larger than μc ($\lambda > \mu c$, or $\varrho > c$) in a delay-system. Then the average number of arriving demands per unit of time (λ) exceeds the total service-rate of c servers (μc). This means an ever-increasing expected number of demands in the queue, unless there is some form of escape from the queue. For example, the queue may be finite in size, then an average of λ–μc arrivals per unit of time find both servers and queue fully occupied and consequently quit; or impatient customers may leave the queue at a sufficient rate (i.e. the queue builds up until the number of impatient demands leaving it reaches an average of λ–μc per unit of time). When, however, no such "safety valve" exists, no steady-state is possible.

When it is not possible for arriving demands to leave the system without being served, the ratio $\eta = \lambda/\mu c = \varrho/c$ is equal to the average rate of occupation of the servers; η will then be called the mean *occupancy* or *efficiency* of the servers.

1.1.5. WHAT ARE THE QUESTIONS THAT ARE NORMALLY ASKED?

The main characteristic of the model described so far is that the sources by far outnumber the servers, which circumstance may result in hindrance of some sort for the sources. In the first place it is therefore necessary to describe a measure for that hindrance; the model must be such that the rate of hindrance, expressed in this measure, can be found as a function of the parameters (e.g. λ, μ and c) governing the performance of the system. The hindrance may, of course, be reduced by increasing the number c of servers. A further aim of an (analytical) theory or treatment of the model may then be to balance the cost of the servers against the "cost" of the hindrance. In order to do so, however, it is primarily necessary to quantify the hindrance.

The most common measures of hindrance are:

(i) In blocking-systems: the *probability of blocking*. This is the proportion of demands that on an average do not receive service. As, in the gas station example, potential customers that do not wait mean a loss of trade, the probability of blocking sometimes is called the *loss-factor*.

(ii) In delay systems (infinite queue): in this case there are at least three plausible measures:

(a) *the probability of delay*, i.e. the average fraction of demands that are not given immediate service;

(b) *the average waiting-time* (either averaged over all demands or over delayed demands only);

(c) *the probability of a waiting-time in excess of some given value*; this is important when perishables are obliged to wait; the waiting results in damage when some critical lapse of time is exceeded.

1.1.6. SITUATIONS NOT COVERED BY THE DESCRIBED MODEL

By combining the various assumptions possible within the model, a multitude of situations arise, a few of which only will be discussed in the sequel. There are, however, some important cases that are not covered by the given model.

In some cases the assumption that any idle server may help any arriving demand is violated. Take a system with three servers as an example. Suppose that, owing to technical restrictions, the sources that produce demands cannot test more than two servers for the "idle" condition. If those two tested servers are both engaged when such a demand is made, this demand is discarded (lost). One way of coping with this difficulty is the following. The group of sources is split up into two subgroups, both having scaled arrival rates $\frac{1}{2}\varrho$. Demands stemming from sources in the first subgroup are allowed to test first server *A* and then server *C*, while those from the second subgroup may test servers *B* and *C* consecutively (cf. Fig. 1.1). The servers *A* and *B* are

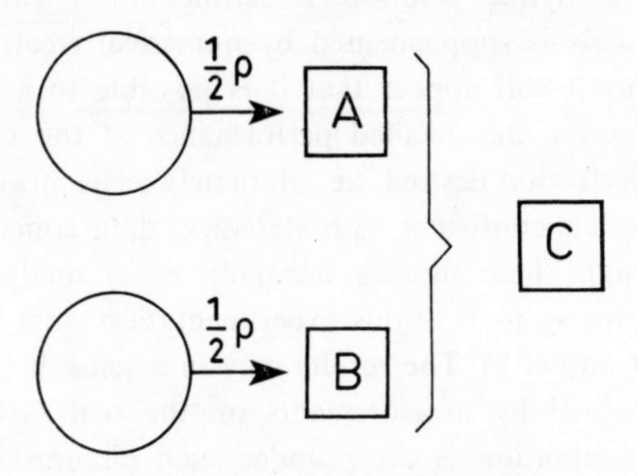

FIG. 1.1. Restricted availability.

called the *individual choices*, server *C* the *common choice*. This case is called *restricted availability* (contrary to the normal case of *full availability*). In telephony many ingenious schemes have been devised with hundreds of servers and as many as forty subgroups, each of which disposes of individual, partially common and completely common choices to a total of, say, twenty choices per subgroup. The way in which the choice of servers for the subgroups has been regulated is called the *grading*. This case will be given some attention in Chapter 7.

Another case that is not covered by the original model occurs when the demands arrive or are served in *batches* (or in *bulk*). Some examples will be dealt with in Chapter 8.

Many of the more complicated cases, as, for example, *simultaneous selection* (in so-called link-systems in telecommunication), will not be dealt with, as they need too specific a knowledge of the technical situations in which those problems arise.

1.1.7. ANALYSIS VERSUS SIMULATION

Up to now there has been a tendency to use analytical tools that become more and more complicated. Sometimes the analysis seems to be a goal in itself, especially so when complicated problems are over-simplified in order to make them fit into some analytical theory. There are, however, better ways of dealing with situations that are too complicated to allow an analytical solution. Experimentation will always be an alternative to analysis supplemented by numerical treatment. Now, in the next sections it will appear that it is possible to set up very fine models that describe the detailed performance of the real systems to any degree of perfection desired, i.e. ultimately as accurate as necessary to make the models consistent with statistical data concerning the real systems. Although these models normally resist analysis, they lend themselves admirably to fictitious experimentation, that is to so-called *simulation* (cf. Chapter 9). The results may in a sense be more accurate than those obtained by measurements on the real systems. For the fictitious experimentation is done under such circumstances that the various parameters that govern the systems' behaviour are controlled much better than is possible with actual experiments. Thus simulation is in some ways a superior kind of laboratory experiment. In the past simulation has not been given the attention it deserved, mostly because it involved a huge amount of very tedious though simple data manipulation. It has now, however, become possible to relegate the burden of this drudgery to the computer, rendering simulation a very fruitful activity. As simulation means measurement, albeit on a mathematical model, the question of the accuracy arises. This will also be dealt with in Chapter 9.

1.2. Characteristics of holding-times; continuation

The holding-time $\underline{h}$ is a so-called *stochastic variable*, i.e. its value cannot be predicted with certainty but only in a statistical way. The prediction is based on statistical information that has been obtained in situations in the past that are supposed to be identical to the present situation in a statistical sense. One way of specifying this statistical knowledge about the holding-time $\underline{h}$ is by giving its *probability density function* (p.d.f.) $f(t)$. This means that $f(t)\,dt$ is the probability that $\underline{h}$

has a value between t and $t + dt$, or, in a notation that will be used in the sequel, that "$\underline{h}$ has a value $t(+dt)$".

The symbol := will be used as a definition symbol. The expression "α is defined as being equal to β" will be denoted by "$\alpha := \beta$". In this shorthand notation we then have

$$f(t)\,dt := Pr\{\underline{h} = t(+dt)\}. \tag{1.2.1}$$

From this definition it follows that

$$Pr\{a < \underline{h} < b\} = \int_a^b f(t)\,dt. \tag{1.2.2}$$

When $a = 0$ and $b = \infty$ this should be equal to unity:

$$\int_0^\infty f(t)\,dt = 1. \tag{1.2.3}$$

Another way of specification is by means of the *cumulative distribution function* (c.d.f.) $F(t)$:

$$F(t) = \int_0^t f(u)\,du = Pr\{\underline{h} \leqq t\}. \tag{1.2.4}$$

Equivalently, we have

$$Pr\{\underline{h} > t\} = 1 - F(t) = \int_t^\infty f(u)\,du.$$

The c.d.f. is zero for $t = 0$, and is non-decreasing, tending to unity for $t \to \infty$ (cf. Fig. 1.2).

The average† holding-time h is given by the first moment:

$$h := \int_0^\infty tf(t)\,dt. \tag{1.2.5}$$

† For some stochastic variables, like $\underline{h}$, the average will be used very much. In such cases the average will be denoted by the same quantity without a bar (h for instance). For other stochastic variables the not-underlined symbol is considered "free". We then may use, for example, $\underline{r} = r$ to express that the stochastical variable $\underline{r}$ has the specific or momentary value r.

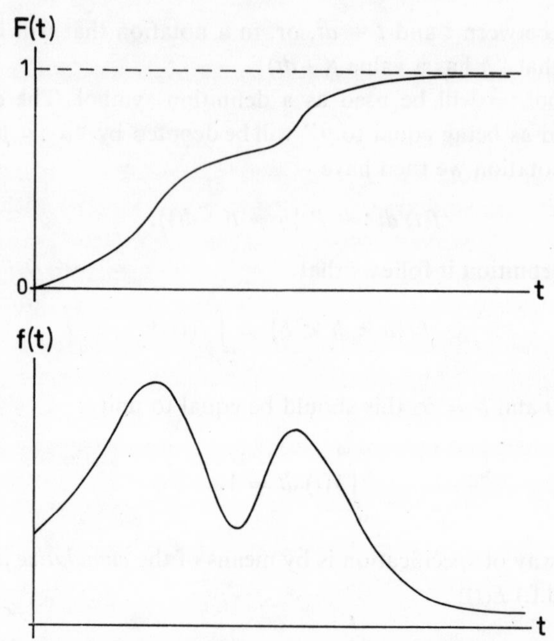

FIG. 1.2. Cumulative distribution function (c.d.f.) and probability density function (p.d.f.) of holding time.

When no specification of the p.d.f. (or c.d.f.) is made, we shall speak about the *general holding-time* case. The only assumption to be made is that the c.d.f. is non-decreasing between 0 and 1. Further restrictions may lead to special classes of c.d.f.'s. The restrictions should preferably be such that:

(i) the restricted class contains c.d.f.'s that cover any available statistical material to a sufficient degree;

(ii) the analysis using a c.d.f. from the restricted class should be simpler than when it is unrestricted.

Usually c.d.f.'s will be supposed to be taken from the *CM*-class, which is defined by the properties:

(*C*) the c.d.f. is continuous;

(*M*) the moments are finite $\left(\int\limits_0^\infty t^r f(t)\, dt < \infty \text{ for positive integral } r\right)$.

The property (M) can be realized in an easy way, e.g. by letting the corresponding p.d.f. be equal to zero for $t > T$, where T is a prescribed upper bound.

Statistical material is always finite in number of items of data and their values. Therefore it is never possible to conclude from any statistical material available that the c.d.f. must necessarily be discontinuous. Moreover, as no observed holding-time can exceed, let us say, 100 years, some suitable upper bound T may always be found. Therefore, there will always be c.d.f.'s in the CM-class that are compatible with any given statistical material. Consequently the CM-class may be said to be completely general. Any attempt to generalize further than this class should be rejected as being "art for art's sake".

A special distribution from the CM-class that is frequently used is the:

Exponential distribution with mean h:

$$f(t) = \frac{1}{h} e^{-t/h},$$
$$F(t) = 1 - e^{-t/h}. \tag{1.2.6}$$

The fact that a quantity has an exponential distribution with mean h will often be described by stating that this quantity "is exponential (h)". Other distributions from the CM-class will be dealt with in due course.

Though generalization further than the CM-class is not necessary, it may occasionally be useful to adopt a *special* c.d.f. not included in the CM-class, such as, for example, the case of:

Constant holding-times:

$$\underline{h} = \text{const} = h,$$
$$F(t) = \begin{cases} 0 & (t < h), \\ 1 & (t \geq h). \end{cases} \tag{1.2.7}$$

In the constant holding-time case the p.d.f. degenerates to an improper function (Dirac function). In some practical cases it is clear that the holding-times are nearly constant. In other cases, however, there is sometimes a surprisingly good agreement with an exponential distribution, e.g. in the case of duration of telephone calls (Erlang, 1925; Molina, 1927). From an analytical point of view the exponential assumption is very appropriate, as will be made clear in the following pages. For this reason this assumption is made whenever reality is not too far away from this distribution.

When an occupation is observed at some instant, this instant divides the holding-time $\underline{h}$ into a *past duration* or *age* p and a *future* or *residual duration* $\underline{r}$. Of course $\underline{h} = p + \underline{r}$. One may now investigate the p.d.f. of the residual duration $\underline{r}$, both unconditional and conditional, subject to the condition $p = u$.

Let $\bar{f}(t)$ be the *unconditional p.d.f. of the residual duration*, i.e. $\bar{f}(t)\,dt := Pr\{\underline{r} = t(+dt)\}$. The residual duration exceeds t if the observed occupation has a holding-time $\underline{h} > t$ and, moreover, the observation is made within a lapse of time $\underline{h} - t$ following its beginning. Hence, $Pr\{\underline{r} > t\}$ equals the expectation of the excess of $\underline{h}$ over t (when positive), divided by the average holding-time:

$$Pr\{\underline{r} > t\} = \int_t^\infty (u - t) f(u)\, du/h.$$

The unconditional p.d.f. of the residual duration is the inverse of the derivative of this expression,

$$\bar{f}(t) = \frac{1}{h}\int_t^\infty f(u)\, du = \frac{1}{h}\{1 - F(t)\} \tag{1.2.8}$$

where $F(t)$ is the c.d.f. of the holding-time $\underline{h}$.

The average residual duration is

$$E(\underline{r}) = \frac{1}{h}\int_0^\infty t\{1 - F(t)\}\frac{dt}{h} = \frac{t^2}{2h}\{1 - F(t)\}\Big|_0^\infty + \frac{1}{2h}\int_0^\infty t^2 f(t)\, dt.$$

For distributions from the CM-class the integrated part vanishes as

$$\lim_{t\to\infty} t^2\{1 - F(t)\} = \lim_{t\to\infty} t^2 \int_t^\infty f(u)\, du \leq \lim_{t\to\infty} \int_t^\infty u^2 f(u)\, du = 0$$

(property M). Hence,

$$E(\underline{r}) = E(\underline{h}^2)/2h. \tag{1.2.9}$$

Now, let $f(t, u)$ be the *conditional p.d.f. of the residual duration* $\underline{r}$, *subject to the condition* $p = u$, where† $f(t, u)\,dt = Pr\{\underline{r} = t(+dt) \mid p = u\}$. The matter of interest is the (conditional) probability that the residual duration $\underline{r}$ should exceed a stated value t, when it is given that the present age is u, that is that the total duration is at least u. We have

$$Pr\{\underline{r} > t \mid \underline{h} > u\} = Pr\{\underline{h} > t + u \mid \underline{h} > u\} = \frac{Pr\{\underline{h} > t + u\}}{Pr\{\underline{h} > u\}}$$

$$= \frac{1 - F(t + u)}{1 - F(u)}. \tag{1.2.10}$$

The corresponding conditional p.d.f. $f(t, u)$ of the residual duration $\underline{r}$, given that the age is u, is obtained by differentiating this expression with respect to t, yielding the result

$$f(t, u) = \frac{f(t + u)}{1 - F(u)}. \tag{1.2.11}$$

Thus $f(t, u)\,dt$ is the probability that an occupation of age u will end after a further duration of between t and $t + dt$. The probability that an occupation of age t will end "immediately", that is within the interval $(t, t + dt)$, is $f(0, t)\,dt$. The quantity

$$g(t) := f(0, t)$$

will be called the (*conditional*) *termination rate for occupations of age t*, or the *age-specific termination rate for age t*. Thus

$$g(t) = \frac{f(t)}{1 - F(t)}. \tag{1.2.12}$$

When the distribution is exponential, $\tilde{f}(t)$, $f(t, u)$ and $f(t)$ all are equal to $e^{-t/h}/h$. Moreover, $g(t) = 1/h$. Verbally: *when the distribution of the holding-time is exponential, the residual duration has the same distribution, both unconditionally and conditionally; the conditional termination rate equals the unconditional termination rate.*

† $Pr\{A \mid B\}$ means the probability of A under the condition B.

This is sometimes called the *property of forgetfulness* (Saaty, 1961). In the early days of life insurance the advantage of the property of forgetfulness has already been recognized, as it renders analysis very simple. It means that, when the exponential distribution holds, all predictions as to the future life chances of all individuals would be independent of age. Therefore premiums could be independent of age. Unfortunately for insurance practice, however, this assumption has proved to be too far from reality.

If at a certain moment r occupations are observed which all possess the same distribution of holding-times and if they have past durations $t_1, \ldots, t_r$, the probability that one of them will end during the next interval Δt is

$$\sum_{i=1}^{r} g(t_i)\,\Delta t \cdot \prod_{j=1,\neq i}^{r} \{1 - g(t_j)\,\Delta t\} \approx \Delta t \sum_{j=1}^{r} g(t_i). \qquad (1.2.13)$$

The quantity $\sum_{i=1}^{r} g(t_i)$ may be termed the *total conditional termination rate.* If the holding-times are exponential(1), this rate reduces to r. In the exponential case knowledge, partial or complete, of the history of occupations does not influence the termination rate.

We shall need some results which are connected with $g(t)$. Let

$$G(t) := \int_0^t g(u)\,du. \qquad (1.2.14)$$

The following relations can easily be verified:

$$g(t) = f(t) \Big/ \int_t^\infty f(u)\,du, \qquad (1.2.15)$$

$$G(t) = -\ln \int_t^\infty f(u)\,du, \qquad (1.2.16)$$

$$e^{-G(t)} = \int_t^\infty f(u)\,du, \qquad (1.2.17)$$

$$\int_0^\infty e^{-G(t)}dt = h. \qquad (1.2.18)$$

From now on we shall normally use time-scaling (cf. Section 1.1.2) by reducing quantities with time dimension by a factor h. The Greek character τ will be reserved to denote "normed times": $\tau = t/h$. The characters u and v will be used to denote both normed and unnormed quantities with time-dimension, but only in places where the context unambiguously shows the meaning. When the p.d.f. and c.d.f. in the normed time notation are given by $f^*(\tau)$ and $F^*(\tau)$, respectively, we get

$$f^*(\tau) = hf(\tau h) \quad \text{and} \quad F^*(\tau) = F(\tau h); \tag{1.2.19}$$

moreover,

$$\int_0^\infty \tau f^*(\tau)\, d\tau = 1. \tag{1.2.20}$$

The asterisks will be dropped when there is no danger of confusion. It should be noted that (1.2.18) now reads

$$\int_0^\infty e^{-G(\tau)} d\tau = 1. \tag{1.2.21}$$

All the other foregoing results remain valid when t is replaced by τ and h by 1. In particular we may note for the case of an exponential distribution with average (1):

(i) the p.d.f. and c.d.f. are given by

$$f(\tau) = e^{-\tau}, \qquad F(\tau) = 1 - e^{-\tau}; \tag{1.2.22}$$

(ii) both the conditional and unconditional termination rate of single occupations are 1;

(iii) the total conditional termination rate of r occupations reduces to r.

1.3. Characteristics of the arrival process

Up to now the arrival process has been vaguely characterized by stating that the arrival rate ϱ should remain constant in some global way. Moreover, it has been assumed that the sources are numerous.

Further, those sources are non-co-operative. A better characterization can be made by making assumptions concerning the mechanism behind the process, i.e. the behaviour of the sources.

1.3.1. THE POISSON ARRIVAL PROCESS

As the sources do not co-operate it is natural to assume that the probability of more than one arrival during $\Delta\tau$ is of the order $\Delta\tau^2$. Hence, neglecting higher-order effects, the probability of one arrival during $\Delta\tau$ equals the expectation of the number of arrivals in $\Delta\tau$, i.e. $\varrho\Delta\tau$. This is the *unconditional* probability, which is valid in the absence of any knowledge. Now, assume that at the beginning of the interval $\Delta\tau$ the history of demands made previously is known (at any rate to some extent). This *may* influence our judgement concerning the probability of an arrival during the interval $\Delta\tau$. For example, let the arrivals refer to the instants at which cars arrive at a certain point of a road. When it is known that those cars have passed some bottle-neck just before, they may be supposed to arrive rather regularly. Hence, when we know that a car has just passed, the conditional probability for an arrival during the next interval $\Delta\tau$ may be much less than $\varrho\Delta\tau$ (even zero) (cf. Fig. 1.3). So there is a negative correlation between arrivals in adjacent intervals. Now, take the case where there is a traffic light just before the mentioned point of the road. When we then know that a car has just passed, our conditional probability for an arrival during the

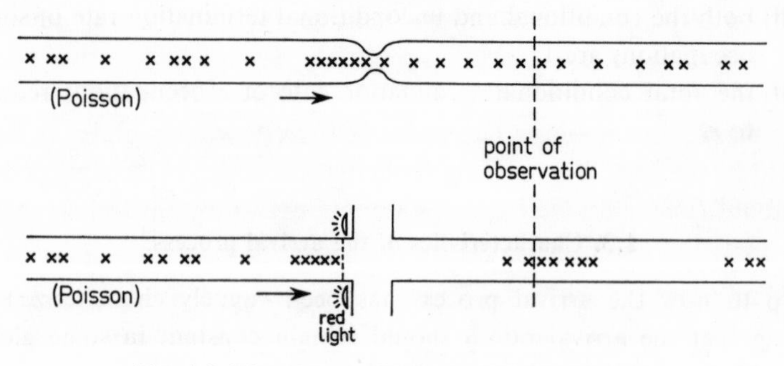

FIG. 1.3. Disturbance of Poisson arrival processes.

next interval $\Delta\tau$ will be *larger* than normal, as we know the cars drive in *clusters*. The correlation between arrivals in adjacent intervals is now positive.

The correlation in these cases is clearly caused by the disturbing influences. When no such influences are known to be present, it is reasonable to put the correlation equal to zero, i.e. assume that history does *not* influence the future. So we assume that the probability of an arrival during $\Delta\tau$ is $\varrho\Delta\tau$, *irrespective of any knowledge about past arrivals.* When this assumption holds, the arrival process is said to be a *Poisson process.* The Poisson arrival process seems to be a natural consequence of: (i) the fact that the sources are very numerous, and (ii) the fact that they are non-co-operative. When at the beginning of the interval $\Delta\tau$ the number of busy servers is known, this constitutes some vague knowledge, so called *inferential knowledge* (Fry, 1928), about past arrivals. Suppose, for example, that no server is busy. It is quite clear that there have been *no* recent arrivals in this case. This constitutes inferential knowledge in the sense referred to. Now, when the arrival process is Poissonian, this (inferential) knowledge does *not* influence the probability of an arrival during the next interval $\Delta\tau$: the arrival rate remains ϱ.

As the probability of one arrival during $\Delta\tau$ is $\varrho\Delta\tau$, and multiple arrivals have probabilities $O(\Delta\tau^2)$, the probability of *no* arrival during $\Delta\tau$ is $1 - \varrho\Delta\tau$.

1.3.2. VIRTUAL AND ACTUAL DEMANDS; TIME- AND DEMAND-AVERAGES

If at a certain moment a demand is made, the past (i.e. *past* demands, *past* durations of occupations) completely determines whether such a demand is lost (in a blocking system) or delayed (in a delay system). Also an average waiting-time, under the circumstances given, is determined by the past. When talking about the probability of blocking or delay or about (unconditional) average waiting-time those quantities can be interpreted in two ways:

(i) as related to the fate of a *virtual demand*, i.e. of a fictitious demand superimposed randomly on a flow of actual demands; they are then called *time-averages*;

(ii) as related to the fate of the *actual demands*; the quantities now are called *demand-averages*.

Demand-averages are the quantities that are of importance to customers. Time-averages are sometimes easier to calculate.

Now, for a Poisson arrival process, those time- and demand-averages are equal. Let $\underline{x}$ be some stochastic quantity—say the length of some queue—that depends on the history of the system (e.g. on the position of *past* demands). Consider a large number N of intervals $\Delta\tau$, chosen at random. The time-average of $\underline{x}$ will be the expectation of the average of the N values of $\underline{x}$ observed at the beginnings of those N intervals. The number of intervals that contain demands will be about $N' = N \cdot \varrho\Delta\tau$. The demand-average of $\underline{x}$ is equal to the expectation of the average of the N' values of $\underline{x}$ observed at the beginnings of those N' intervals. Now the Poisson process guarantees that the picking out of the N' intervals from the group of N is carried out independently of the history and hence independently of the values of $\underline{x}$ at the beginnings of those intervals. This then proves that time- and demand-averages are equal, in case the arrivals form a Poisson process.

1.3.3. A SIMPLE BIRTH-PROCESS; THE POISSON DISTRIBUTION

Suppose we wish to evaluate the probability $p_r(\tau)$ of having exactly r arrivals of a Poisson process in an interval of length τ. Consider the interval $I = (0, \tau + \Delta\tau)$ and let it be split up into $I_1 = (0, \tau)$ and $I_2 = (\tau, \tau + \Delta\tau)$. The event "$r$ arrivals in I" can be caused by any of $r + 1$ compound events:

(0) "r arrivals in I_1" *and* "no arrivals in I_2"

or

(1) "$r - 1$ arrivals in I_1" *and* "one arrival in I_2"

or

(2) "$r - 2$ arrivals in I_1" *and* "two arrivals in I_2"

or

. . . .

or

(r) "no arrivals in I_1" *and* "r arrivals in I_2".

For a Poisson process the left- and right-hand events of each of the $r + 1$ compounds are independent. Hence, the probabilities of the compound events are the products of the unconditional probabilities of their constituents. Now, the probabilities of two or more arrivals in $I_2 = (\tau, \tau + d\tau)$ are $O(\Delta\tau^2)$ or higher. Hence, only the compound events (0) and (1) have non-negligible contributions. So, for $r = 0, 1, 2, \ldots$,

$$p_r(\tau + \Delta\tau) = p_r(\tau) \cdot (1 - \varrho\Delta\tau) + p_{r-1}(\tau) \cdot \varrho\Delta\tau + O(\Delta\tau^2)$$

or

$$[p_r(\tau + \Delta\tau) - p_r(\tau)]/\Delta\tau = \varrho p_{r-1}(\tau) - \varrho p_r(\tau) + O(\Delta\tau).$$

As -1 arrivals in I_1 is impossible, p_{-1} should be given the value 0. Passing to the limit $\Delta\tau \to 0$ and dropping the argument τ:

$$\frac{dp_r}{d\tau} + \varrho p_r = \varrho p_{r-1} \quad (r = 0, 1, 2, \ldots; p_{-1} \equiv 0). \tag{1.3.1}$$

The initial conditions are

$$p_r(0) = \begin{cases} 1 & (r = 0), \\ 0 & (r > 0). \end{cases} \tag{1.3.2}$$

In order to integrate this system of differential equations let us put $p_r(\tau) = e^{-\varrho\tau}k_r(\tau)$. Insertion in (1.3.1) and (1.3.2) yields

$$\frac{dk_r}{d\tau} = \varrho k_{r-1}, \tag{1.3.3}$$

$$k_r(0) = \begin{cases} 1 & (r = 0), \\ 0 & (r > 0) \end{cases} \tag{1.3.4}$$

and hence,

$$\frac{dk_0}{d\tau} = 0, \qquad k_0(0) = 1 \Rightarrow k_0(\tau) = 1,$$

$$\frac{dk_1}{d\tau} = \varrho \cdot 1, \quad k_1(0) = 0 \Rightarrow k_1(\tau) = \frac{\varrho\tau}{1!},$$

$$\frac{dk_2}{d\tau} = \varrho \cdot \frac{\varrho\tau}{1!}, \quad k_2(0) = 0 \Rightarrow k_2(\tau) = \frac{\varrho^2\tau^2}{2!}.$$

Continuing in this way one obtains

$$k_r(\tau) = (\varrho\tau)^r/r!$$

$$p_r(\tau) = e^{-\varrho\tau}(\varrho\tau)^r/r! \tag{1.3.5}$$

These $p_r(\tau)$ constitute a *discrete distribution function*, which is the well-known *Poisson distribution*. For $\varrho\tau \gg 1$ it has a peak near $r \approx \varrho\tau$, which is the more pronounced the larger $\varrho\tau$ is. For $\varrho\tau \to \infty$ it tends to a normal distribution.

In the general system described in Section 1.1 the arrivals may be considered as items that "enter a system" or that are "born in the system". Those items then have a "life" within the system as "waiting demands" and/or "occupations", after which they "leave the system" or "die". The system will be said to be in "state r"—or $[r]$ for short—when there are r items in the system. In the case described in this section there are only items entering the system, they do not leave.

Now, consider a large number (N) of those systems, all beginning in $[0]$ at time 0. At any time τ the increase per unit of time of systems in $[r]$—i.e. $d/d\tau(Np_r)$—should be equal to the number of systems that assume the $[r]$ state per unit of time—i.e. ϱNp_{r-1}—minus the number ϱNp_r of systems that leave the $[r]$ state. So,

$$\frac{d}{d\tau}(Np_r) = \varrho Np_{r-1} - \varrho Np_r. \tag{1.3.6}$$

This supplies another explanation of (1.3.1). The equations (1.3.1) are the so-called *birth-equations*.

1.3.4. THE DISTRIBUTION OF INTERARRIVAL-TIMES FOR THE POISSON ARRIVAL PROCESS

The lapse of time between two consecutive arrivals will be termed an *interarrival-time* $\underline{i}$ (a stochastic variable). Its c.d.f. will be denoted by $A(\tau)$, its p.d.f. by $a(\tau)$:

$$\left.\begin{aligned} A(\tau) &:= Pr\{\underline{i} \leqq \tau\}, \\ a(\tau)\,d\tau &:= Pr\{\underline{i} = \tau(+d\tau)\} = A(\tau)\,d\tau. \end{aligned}\right\} \tag{1.3.7}$$

Now,

$$Pr\{\underline{i} < \tau\} = 1 - Pr\{\text{no arrival (in } 0, \tau) \mid \text{last arrival at } 0^-\}. \quad (1.3.8)$$

The condition is not relevant for a Poisson arrival process, as the past does not influence the future. Hence, the latter probability in (1.3.8) equals $p_0(\tau)$, introduced in Section 1.3.3. So by using (1.3.5) one obtains

$$A(\tau) = 1 - e^{-\varrho\tau} \quad (1.3.9)$$

and hence,

$$a(\tau) = \varrho e^{-\varrho\tau}. \quad (1.3.10)$$

Comparison with (1.2.6) shows that the interarrival-times are exponential $(1/\varrho)$. The mean interarrival-time is $1/\varrho$, as could be expected. The Poisson arrival process is, however, not completely determined by the distribution function of interarrival-times. This should be supplemented by the statement that successive interarrival-times are independent (as in a Poisson process past and future are independent).

The lapse of time between an arbitrary instant and the next arrival will be called the *residual interarrival-time*. According to the property of forgetfulness (Section 1.2) its distribution also is exponential $(1/\varrho)$.

1.3.5. OTHER DISTRIBUTIONS OF INTERARRIVAL-TIMES

As far as *holding-times* are concerned, the exponential distribution is just one distribution. Other distributions may be equally plausible. Hence, whenever it is possible to tackle a problem in an analytical way under the assumption of an arbitrary distribution function of holding-times (e.g. in the *CM*-class), this should be preferred.

There also have been many investigations using an arbitrary distribution of interarrival-times. In the opinion of the present author this is, however, not of the same practical importance. It has been pointed out earlier that a multitude of non-co-operative sources produce arrivals that naturally tend to a Poisson distribution. Departures from this distribution can be brought about by external circumstances that cause delays of arrivals (cf. the examples of cars that pass through bottle-necks or that are subject to traffic regulations; Fig. 1.3). But when those causes influence interarrival-times, they disturb the freedom from correlation of successive interarrival-times as well! As the investigations referred to with arbitrary interarrival-time distribution do *not* cope with correlation,† they do not seem to cover the general case. For the present the Poisson process will be considered the only possibility of a description of an arrival process that is both simple enough and realistic.

When the departure from the Poisson process is serious, one should cope with it in such a way that the disturbing mechanism is considered a part of the system described. This will most certainly render the model so complex as to exclude analytical treatment. As will be pointed out in Chapter 9 there are, however, other tools in this case.

A meaningful variant to the Poisson process can be obtained by supposing that the arrival rate ϱ is not constant but varies with time "very slowly" (Palm, 1943). Still another variant is formed by assuming that the number of sources is limited N. When at a certain time r servers (and consequently r sources) are engaged, it is realistic to take the arrival rate proportional to the number of free sources, i.e. $\alpha(N - r)$ (Engset, 1918).

1.4. Kendall's classification

D. G. Kendall (1951) has introduced the following classification of the problems at hand. Three symbols are given, separated by two oblique strokes: —/—/—. The last symbol specifies the number of servers. The first and second symbols denote the type of distribution of the interarrival-time and of the holding-time, respectively. An exponential distribution is indicated by the symbol M (called after the "Markovian property", which essentially is the property of forgetfulness). Constant intervals are characterized by the symbol D (deterministic). Arbitrary distributions are given the symbol G ("general"). If the latter case refers to interarrival-times and if, moreover, successive interarrival-

† The assumption is termed GI ("General Independent") in Kendall's classification (cf. Section 1.4).

times are supposed to be independent, this is indicated by the symbol *GI* ("general independent"). Some examples:

$$M/M/c \text{ means } \begin{cases} \text{Poisson arrival process,} \\ \text{exponential distribution of holding-times,} \\ c \text{ servers;} \end{cases}$$

$$M/G/1 \text{ means } \begin{cases} \text{Poisson arrival process,} \\ \text{arbitrary distribution of holding-times,} \\ \text{one server;} \end{cases}$$

$$GI/D/1 \text{ means } \begin{cases} \text{interarrival-times ``general independent'',} \\ \text{constant holding-time,} \\ \text{one server.} \end{cases}$$

In the cases to be dealt with, the first symbol will nearly always be *M* (Poisson). Blocking and delay systems will be distinguished by simply writing "blocking" or "delay" behind the classification.

2. THE STANDARD CASES

We shall term "standard cases" the simple steady-state cases $M/M/c$-blocking and $M/M/c$-delay. The arrival process is Poissonian in both cases and the distribution of holding-times is exponential. Moreover, it is assumed that the order of serving delayed demands is *not* dependent on *a priori* knowledge of holding-times.

2.1. Probability of blocking for the system $M/M/c$-blocking

The arrival process is Poissonian with arrival (or birth) rate ϱ. There are c servers. The holding-time is exponential (1). There is no waiting facility. If all servers are occupied, arriving demands are discarded.

When the number of busy servers is r ($r = 0, 1, \ldots, c$) the system will be said to be "in state $[r]$". It is assumed that stationarity exists. This means that the probability of the system being in $[r]$ is time-independent:

$$p_r := Pr\{[r] \text{ at time } \tau\} = \text{const.} \tag{2.1.1}$$

At an arrival (rate $= \varrho$) the index r of the state of the system is increased by 1 as long as $r < c$. If $r = c$ an arrival is ineffective. When the system is in $[r]$ the arrival rate remains ϱ, as for the Poisson process the inferential knowledge conveyed to us by the specification of the state is irrelevant (cf. Section 1.3.1). The total termination rate is r; possible inferential knowledge about the ages of the r occupations is irrelevant as the holding-time is exponential (cf. Section 1.2, property of forgetfulness). The probability of state and rates of arrival and termination (collectively called *transitions*) may be summarized in the following scheme:

states:

$$[0] \underset{1}{\overset{\varrho}{\rightleftarrows}} [1] \underset{2}{\overset{\varrho}{\rightleftarrows}} \cdots [r-1] \underset{r}{\overset{\varrho}{\rightleftarrows}} [r] \underset{r+1}{\overset{\varrho}{\rightleftarrows}} [r+1] \cdots [c-1] \underset{c}{\overset{\varrho}{\rightleftarrows}} [c],$$

probability of states:

p_0 p_1 p_{r-1} p_r p_{r+1} p_{c-1} p_c.

Now, stationarity requires that the probability of the formation of $[r]$ in $\Delta\tau$ should equal the probability of a departure from $[r]$ during $\Delta\tau$. The formation is possible from state $[r-1]$, which has probability p_{r-1}, by an arrival, which has probability $\varrho\Delta\tau$. The joint probability of these events is $p_{r-1} \cdot \varrho\Delta\tau$. When $r < c$, another possibility is from $[r+1]$ (probability p_{r+1}) by a death (conditional probability $(r+1)\Delta\tau$): joint probability $p_{r+1} \cdot (r+1)\Delta\tau$. Formations by multiple transitions have probabilities of order $O(\Delta\tau^2)$ and are discarded. When the system is in $[r]$, where $r < c$, both arrival and termination would mean departure from $[r]$; hence the probability of a departure from $[r]$ in $\Delta\tau$ is: $p_r \cdot (\varrho + r)\Delta\tau$. So

$$p_{r-1} \cdot \varrho\Delta\tau + p_{r+1} \cdot (r+1)\Delta\tau = p_r \cdot (\varrho + r)\Delta\tau$$

or

$$\varrho p_{r-1} - (\varrho + r)p_r + (r+1)p_{r+1} = 0 \quad (r = 0, 1, \ldots, c-1). \qquad (2.1.2)$$

For this equation to hold for $r = 0$ too, p_{-1} should be taken equal to 0. For the state $[c]$ it should be observed that no transitions to and from a state $[c+1]$ are possible. Hence,

$$\varrho p_{c-1} - cp_c = 0. \qquad (2.1.3)$$

The set of eqns. (2.1.2) and (2.1.3) are called the *birth-and-death equations*. It is a set of $c+1$ homogeneous linear equations in $c+1$ unknowns. So, for a non-trivial solution to exist, the equations must be linearly dependent. By adding eqns. (2.1.2) for $0, 1, \ldots, r$ one obtains

$$-\varrho p_r + (r+1)p_{r+1} = 0 \quad (r = 0, 1, \ldots, c-1), \qquad (2.1.4)$$

the last of which is identical to (2.1.3), showing the dependency. The set of eqns. (2.1.4) can be obtained in a straightforward way. If the set of states is partitioned into $A = \{[0], ..., [r]\}$, $B = \{[r + 1], ..., [c]\}$, transitions from A to B and back are identical to transitions from $[r]$ to $[r + 1]$ and back. Now, stationarity requires that the rate of transitions $A \to B$ should be equal to the rate of transitions $B \to A$. This yields:

$$\varrho p_r = (r + 1) p_{r+1}.$$

The set of eqns. (2.1.4) should be supplemented by the statement that the defined states form a complete set of mutually exclusive events, and hence,

$$\sum_{r=0}^{c} p_r = 1. \tag{2.1.5}$$

The eqns. (2.1.4) taken consecutively for $r = 0, 1, 2, ...$ enable one to express p_1 in terms of p_0, p_2 in terms of $p_1, ..., p_c$ in terms of p_{c-1}. Hence, all unknowns can be expressed in terms of p_0:

$$p_r = \frac{\varrho^r}{r!} p_0 \quad (r = 1, ..., c). \tag{2.1.6}$$

By insertion into (2.1.5) one obtains

$$p_0 = 1 \Bigg/ \left\{1 + \frac{\varrho}{1!} + \frac{\varrho^2}{2!} + \cdots + \frac{\varrho^c}{c!}\right\} \tag{2.1.7}$$

and hence,

$$p_r = \frac{p^r}{r!} \Bigg/ \sum_{r=0}^{c} \frac{\varrho^r}{r!}. \tag{2.1.8}$$

The *probability of blocking* B is the probability of finding the system in $[c]$; hence,

$$B = \frac{\varrho^c}{c!} \Bigg/ \sum_{r=0}^{c} \frac{\varrho^r}{r!} (= E_{1,c}(\varrho)). \tag{2.1.9}$$

This is the famous *Erlang B-formula*. The function $E_{1,c}$ is called the *first Erlang function*.

When $c - \varrho > 2\sqrt{\varrho}$ the denominator in (2.1.9) is practically equal to e^{ϱ} and B is nearly equal to the Poisson expression, to be denoted by φ:

$$B \approx \varphi_c := \frac{e^{-\varrho}\varrho^c}{c!}. \tag{2.1.10}$$

Using upper index 1 for "first sums" of arithmetic functions (cf. A 7)† the eqn. (2.1.9) may also be written as:

$$B = \varphi_c / \varphi_c^1. \tag{2.1.9'}$$

In Fig. 2.1 the relation between $E_{1,c}$, ϱ and c is shown. It can be used for dimensioning purposes. When, for example, $\varrho = 3$ and we want B to be not larger than 0.01, the graph shows that c should be $\geqq 8$. So one takes $c = 8$.

The eqns. (2.1.2) and (2.1.3) are called birth-and-death equations because they are connected with systems in which items are born and die. The name stems from biological applications in connection with the growth of colonies of bacteria. The equations may also be interpreted as follows. Consider a large number of identical systems of the type described that are contained in a *hypersystem*. The equations

$$\varrho p_{r-1} + (r+1)\, p_{r+1} = (\varrho + r)\, p_r \tag{2.1.2a}$$

and

$$\varrho p_{c-1} = c p_c \tag{2.1.3a}$$

state that the rate at which systems assume the [r] (or [c]) state within the hypersystem equals the rate at which they leave that state. Thus they may be said to describe the state of *statistical equibrium*, a notion from statistical mechanics. For a more rigorous derivation of the birth-and-death equations recourse should be had to the so-called *Chapman-Kolmogorov equations* (cf. Syski, 1960). More complicated situations may lead to sets of equations that are related not simply to the number of simultaneous occupations but also to attributes of those occupations, e.g. their past durations (cf. Section 1.2). In those cases we shall call these equations *equations of state*.

† A 7 refers to formula A 7 of Appendix.

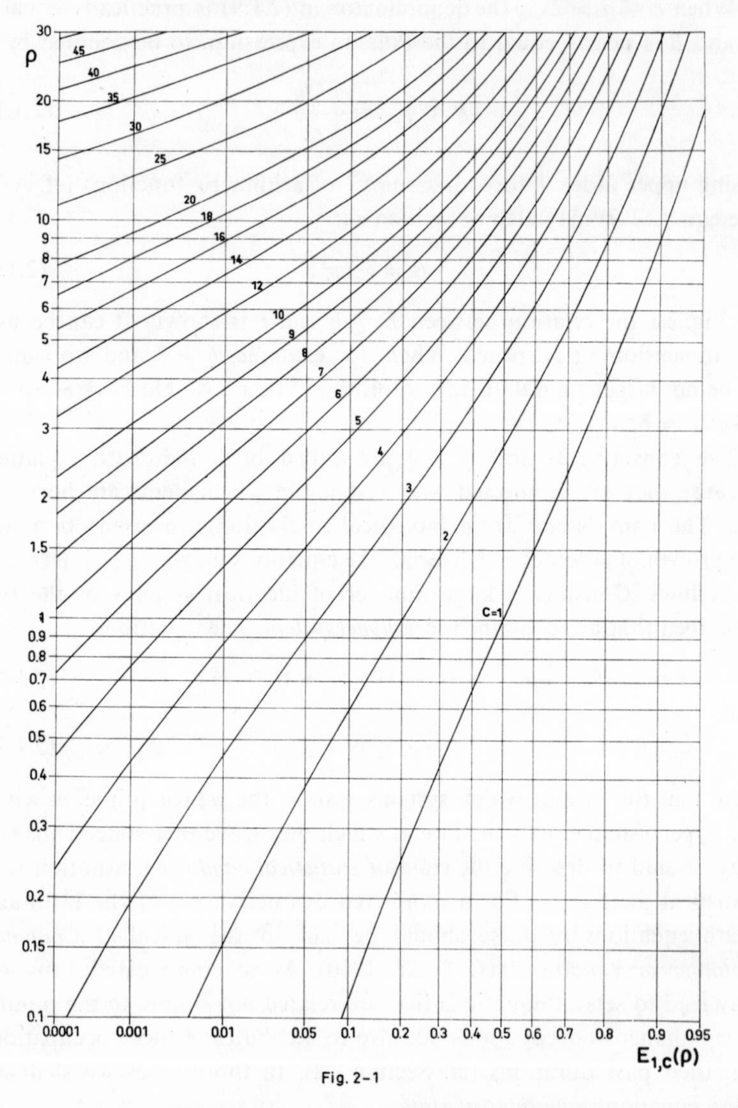

FIG. 2.1. The first Erlang function $E_{1,c}(\varrho)$.

The *offered traffic* ϱ (average number of demands per unit of time) may be split up into the *traffic handled* $(1 - B)\varrho$ (i.e. the average number of demands served per unit of time) and the *lost traffic* $B\varrho$ (average number of lost demands per unit of time). They are all said to be measured in erlangs. The average number of demands that could possibly be handled per unit of time by a group of c servers is c, its *capacity*. Hence, $\eta := (1 - B)\,\varrho/c$ may be called the *overall server efficiency or occupancy*. For constant B it increases with increasing ϱ (cf. Fig. 2.2).

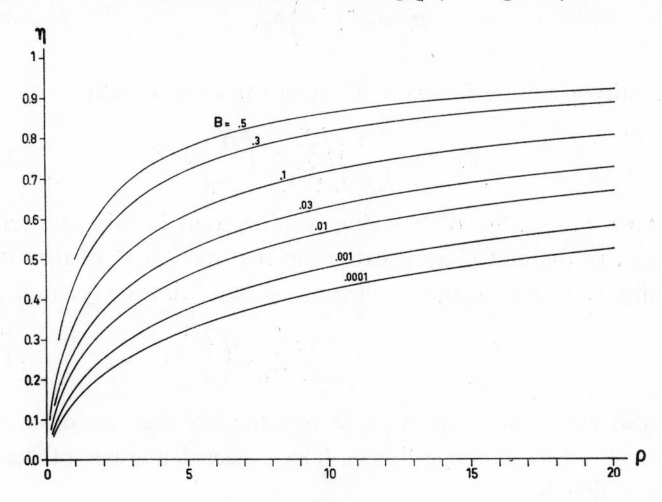

FIG. 2.2. Overall server efficiency.

2.2. The Engset formula

When the number N of sources is small, the arrival rate may be put proportional to the number of free sources: $\alpha(N - r)$. So now the set of states and the probabilities of transition are

$$\underbrace{[0] \underset{1}{\overset{\alpha N}{\rightleftarrows}} [1] \underset{2}{\overset{\alpha(N-1)}{\rightleftarrows}} [2] \ldots [r]}_{A} \underset{r+1}{\overset{\alpha(N-r)}{\rightleftarrows}} \underbrace{[r+1] \ldots \underset{c}{\overset{\alpha(N-c+1)}{\rightleftarrows}} [c]}_{B}.$$

Again, the expected numbers of transitions $A \to B$ and $B \to A$ per unit of time should be equal. Hence,

$$\alpha(N-r)\,p_r = (r+1)\,p_{r+1} \quad (r = 0, 1, \ldots, c-1) \tag{2.2.1}$$

together with

$$\sum_{r=0}^{c} p_r = 1. \tag{2.2.2}$$

In the same way as in Section 2.1 one now obtains

$$p_r = \alpha^r \binom{N}{r} p_0. \tag{2.2.3}$$

The quantity p_0 ensues from the norming equation. Finally,

$$p_r = \alpha^r \binom{N}{r} \Big/ \sum_{r=0}^{c} \alpha^r \binom{N}{r}. \tag{2.2.4}$$

The term probability of blocking is ambiguous in this case (cf. Section 1.3.2). In the first place it can mean the fraction B^t of time during which all servers are engaged (sometimes called "time-congestion")

$$B^t = p_c = \alpha^c \binom{N}{c} \Big/ \sum_{r=0}^{c} \alpha^r \binom{N}{r}. \tag{2.2.5}$$

It can also mean the proportion B^d of demands that are unsuccessful ("demand-" or "call-congestion"). The expected number of demands per unit of time is

$$\sum_{r=0}^{c} p_r \alpha(N-r) = \alpha N \sum_{r=0}^{c} \alpha^r \binom{N-1}{r} \Big/ \sum_{r=0}^{c} \alpha^r \binom{N}{r},$$

and the expectation of unsuccessful demands per unit of time:

$$p_c \alpha(N-c) = \alpha N \cdot \alpha^c \binom{N-1}{c} \Big/ \sum_{r=0}^{c} \alpha^r \binom{N}{r}.$$

The ratio of the second of these to the first is

$$B^d = \alpha^c \binom{N-1}{c} \Big/ \sum_{r=0}^{c} \alpha^r \binom{N-1}{r}. \tag{2.2.6}$$

These expressions for B^t and B^d—the *Engset formulae*—are generally unequal. For $N \to \infty$ they both tend to the Erlang B-formula. They are not used very frequently. In the first place there is an extra parameter, which makes tabulation cumbersome. Moreover, when N is not too large, there seem to be other assumptions that no longer hold; for example, the assumption that unsuccessful demands (in a blocking system) are lost without further consequences. In most cases those lost demands will cause repetition shortly afterwards. In the case of a multitude of sources those repetitions will probably merge into the mass of demands. When N is smaller, however, the repetitions may cause a relatively considerable increase of arrivals during and shortly after a blocking period. In this case any theoretical result that does not take into consideration this repetition effect should be considered suspect.

2.3. Probability of delay and average waiting-time for the system *M/M/c*-delay

Again the arrival process is Poissonian with arrival rate ϱ. There are c servers. The distribution of holding-times is exponential (1). There is an infinite queue.

When there are r demands that have not yet been completely served, the system is said to be "in state $[r]$". When $r \leqq c$, there are r engaged servers, whilst the queue is empty. If $r = c + m$ $(m > 0)$, all servers are engaged, whilst there are m queued demands. It is assumed that there is stationarity. The probability of the system being in $[r]$ is supposed to be time-independent and is denoted by p_r.

When the system is in $[r]$, there is a transition rate ϱ to $[r + 1]$. The transition rate to $[r - 1]$ equals the number of engaged servers, i.e. r when $r \leqq c$ and c otherwise. Hence, the states and the transition rates may be given as follows:

states:

$$[0] \underset{1}{\overset{\varrho}{\rightleftarrows}} [1] \underset{2}{\overset{\varrho}{\rightleftarrows}} [2] \ldots \underset{c-1}{\overset{\varrho}{\rightleftarrows}} [c-1] \underset{c}{\overset{\varrho}{\rightleftarrows}} [c] \underset{c}{\overset{\varrho}{\rightleftarrows}} [c+1] \underset{c}{\overset{\varrho}{\rightleftarrows}} [c+2] \ldots$$

probability of states:

$$p_0 \quad p_1 \quad p_2 \quad \ldots \quad p_{c-1} \quad p_c \quad p_{c+1} \quad p_{c+2} \quad \ldots$$

Just as in Sections 2.1 and 2.2 the probability of a transition $[r] \to [r+1]$ should equal the probability of a reverse transition. Hence,

$$-\varrho p_r + (r+1)\, p_{r+1} = 0 \quad (r = 0, 1, \ldots, c-1) \tag{2.3.1}$$

and

$$-\varrho p_r + c p_{r+1} = 0 \quad (r = c, c+1, \ldots). \tag{2.3.2}$$

Those equations should be completed by

$$\sum_{r=0}^{\infty} p_r = 1. \tag{2.3.3}$$

The eqns. (2.3.1) and (2.3.2) are suitable for expressing p_{r+1} in terms of p_r, this again in terms of $p_{r-1}, \ldots$, and ultimately in terms of p_0. The result is

$$p_r = \begin{cases} \dfrac{\varrho^r}{r!} p_0 & (r = 1, 2, \ldots, c). \\ \left(\dfrac{\varrho}{c}\right)^{r-c} p_c = \left(\dfrac{\varrho}{c}\right)^{r-c} \dfrac{\varrho^c}{c!} p_0 & (r > c). \end{cases} \tag{2.3.4}$$

When $\varrho \geqq c$ and $p_0 > 0$, the array $p_0, p_1, \ldots$ would be never-decreasing. This is incompatible with (2.3.3). When $p_0 = 0$, all p_r would be zero, which does not constitute a good solution. Hence, for $\varrho \geqq c$ no solution is found. This stems from the fact that for $\varrho \geqq c$ stationarity is impossible (Section 1.1.4). For the present we assume $\varrho < c$. When the values (2.3.4) for the p_r are inserted into (2.3.3), the value for p_0 follows from

$$p_0 \left[1 + \frac{\varrho}{1!} + \frac{\varrho^2}{2!} + \cdots + \frac{\varrho^{c-1}}{(c-1)!} + \frac{\varrho^c}{c!} \left(1 + \frac{\varrho}{c} + \frac{\varrho^2}{c^2} + \ldots \right) \right] = 1,$$

yielding

$$p_0 = 1 \Bigg/ \left[1 + \frac{\varrho}{1!} + \cdots \frac{\varrho^{c-1}}{(c-1)!} + \frac{\varrho^c}{c!} \frac{c}{c-\varrho} \right]. \tag{2.3.5}$$

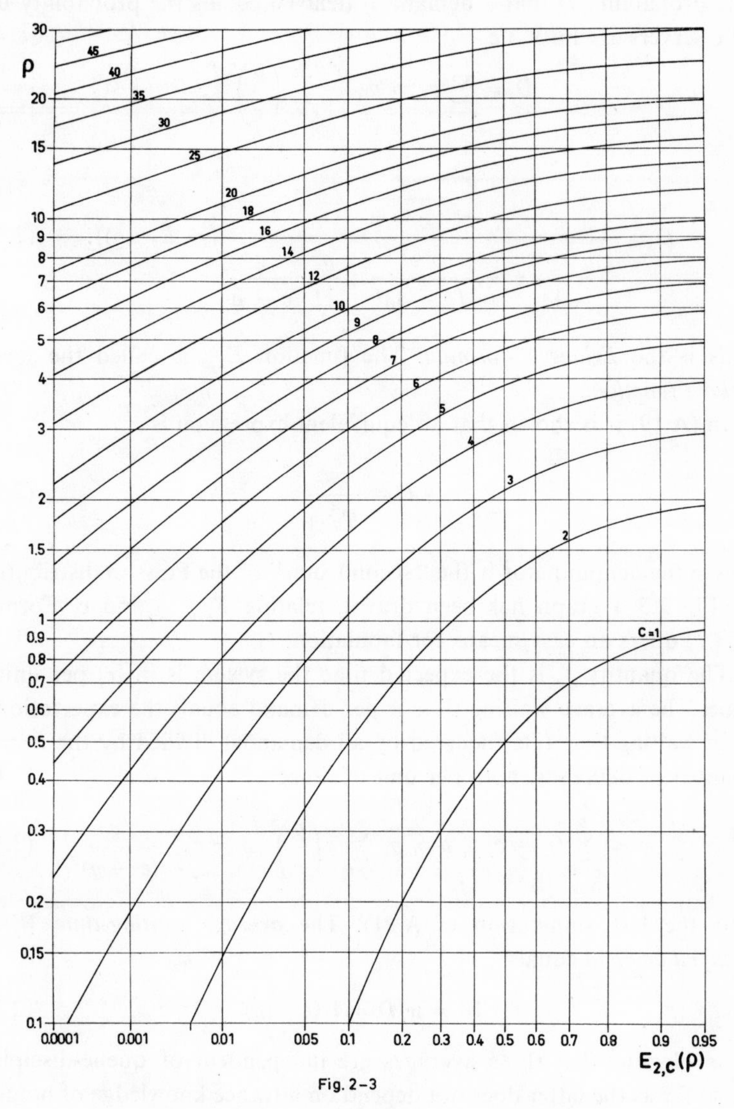

FIG. 2.3. The second Erlang function $E_{2,c}(\varrho)$.

The probability D that a demand is delayed equals the probability that all c servers are busy, i.e.

$$D = \sum_{r=c}^{\infty} p_r = p_0 \frac{\varrho^c}{c!} \sum_{r=c}^{\infty} \left(\frac{\varrho}{c}\right)^{r-c},$$

or

$$D = \frac{\dfrac{\varrho^c}{c!} \cdot \dfrac{c}{c-\varrho}}{1 + \dfrac{\varrho}{1!} + \cdots \dfrac{\varrho^{c-1}}{(c-1)!} + \dfrac{\varrho^c}{c!} \cdot \dfrac{c}{c-\varrho}} \; (= E_{2,c}(\varrho)). \tag{2.3.6}$$

This is the *Erlang C-formula.* The function $E_{2,c}$ is called the *second Erlang function.*

In (A 19) it is shown that an equivalent expression is

$$D = \frac{c\varphi_c}{\varphi_{c-1}^2} \tag{2.3.7}$$

where the denominator is the "second sum" of the Poisson distribution. In Fig. 2.3 a graph has been drawn, relating $E_{2,c}$, ϱ and c. Formula (2.3.7) offers an easy means for tabulation.

The quantity p_r is the expected time the system is in [r] per unit of time. The average waiting-time w per demand equals the expectation of *total* waiting-time (i.e. incurred by all demands) divided by the average number of demands, both per unit of time:

$$w = \sum_{r=c}^{\infty} (r-c)\, p_r/\varrho = p_c \sum_{s=0}^{\infty} s \left(\frac{\varrho}{c}\right)^s \Big/ \varrho = p_c \cdot \frac{c}{(c-\varrho)^2} \tag{2.3.8}$$

(for the last summation cf. A 21). The *average waiting-time W per delayed demand* equals

$$W = w/D = 1/(c-\varrho). \tag{2.3.9}$$

It is obvious that these averages are independent of queue-discipline, in so far as the latter does not depend on advance knowledge of holding-times of waiting demands. The *distribution* of waiting-times, however, is *not* independent of this discipline (cf. Chapter 3).

2.4. Econometric considerations

Given a certain flow of demands and supposing the holding-times to be exponential (1), the formulae derived up to now may be used to answer questions like the following:

(i) How many servers are necessary in a blocking system (delay system) to ensure that not more than a fraction p of all demands is lost (delayed)?

(ii) How many servers are necessary in a delay system to make sure that the average waiting-time does not exceed a certain value T?

The actual values of p and T in these questions, however, are largely arbitrary; they depend on intuition. Under certain circumstances, however, a more rational approach will prove to be possible.

Consider a blocking system with $c - 1$ servers. Its lost traffic (cf. Section 2.1) is $\varrho E_{1,c-1}(\varrho)$. When c servers are taken instead, the lost traffic reduces to $\varrho E_{1,c}(\varrho)$. The difference, say η_c, is to be attributed to the additional cth server. Sometimes, there is a *hunting order* among the servers. This means that server i cannot be seized by an arriving demand unless the servers $1, 2, \ldots, i - 1$ are all engaged. In this case the quantity η_c referred to equals the traffic handled by the cth server (in erlangs). As one server can handle one erlang, η_c is the *efficiency of the c-th server in the hunting order*:

$$\eta_c = \varrho\{E_{1,c-1}(\varrho) - E_{1,c}(\varrho)\}. \tag{2.4.1}$$

In Fig. 2.4 the relation between η_c, ϱ and c is shown.

Suppose the following figures to be known:

k_1 = average profit per demand served;

k_2 = cost of one server, taken per unit of time.

If a server were busy for 100%, it would on the average handle one demand per unit of time and hence make a profit of k_1 per unit of time. Now, the (additional) cth server is busy during a fraction η_c of total

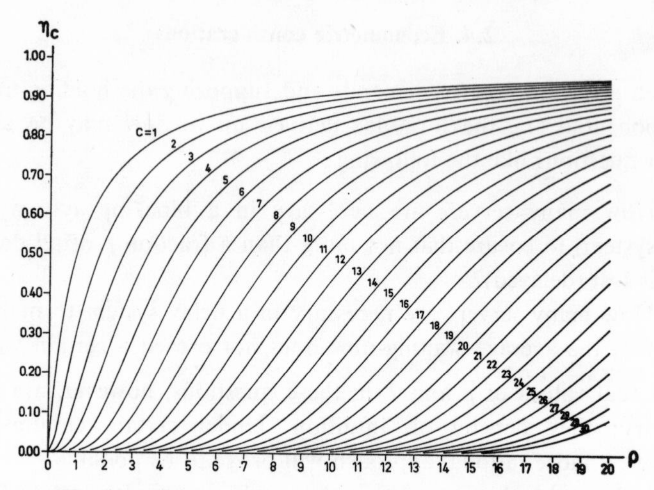

FIG. 2.4. Efficiency of the cth server in a blocking system.

time. Hence its average profit is $\eta_c k_1$ per unit of time. If this is larger than the cost k_2, the cth server is economical, and otherwise not:

$$\text{the } c\text{th server is} \begin{cases} \text{economical} & \text{if } \eta_c > k_2/k_1, \\ \text{not economical} & \text{if } \eta_c < k_2/k_1. \end{cases} \tag{2.4.2}$$

Let $\varrho = 8$ and $k_2/k_1 = 0.45$. From Fig. 2.4 it follows that the efficiencies of servers no. 1, 2, ..., 9 are 0.89, 0.87, ..., 0.50, whilst that of server no. 10 is 0.41. Hence the 10th server is the first in order that ceases to be economical. So the optimal number of servers is 9.

Now, let us turn to the case of a delay-system with c servers ($c > \varrho$, infinite queue). Assume that we know:

k_1 = cost of one demand waiting for one unit of time;

k_2 = cost of one server per unit of time.

The total expected waiting-time per unit of time is $\varrho E_{2,c}/(c - \varrho)$. When $c < \varrho$, the cth server is *necessary*. If not, it is *profitable* provided

$$\frac{\varrho E_{2,c}}{c - \varrho} k_1 + c k_2 < \frac{\varrho E_{2,c-1}}{c - \varrho - 1} k_1 + (c - 1) k_2.$$

Thus,

> the cth server is economical if $c \leqq \varrho$
>
> or else if
>
> $$\varrho \left\{ \frac{E_{2,c-1}}{c - \varrho - 1} - \frac{E_{2,c}}{c - \varrho} \right\} \geqq k_2/k_1 . \tag{2.4.3}$$

When in (2.4.3) the equality sign holds, it makes no difference whether there are $c - 1$ or c servers. In Fig. 2.5 a graph has been drawn, showing k_2/k_1 as a function of ϱ, defined by

$$k_2/k_1 = \varrho \left\{ \frac{E_{2,c-1}}{c - \varrho - 1} - \frac{E_{2,c}}{c - \varrho} \right\} \tag{2.4.4}$$

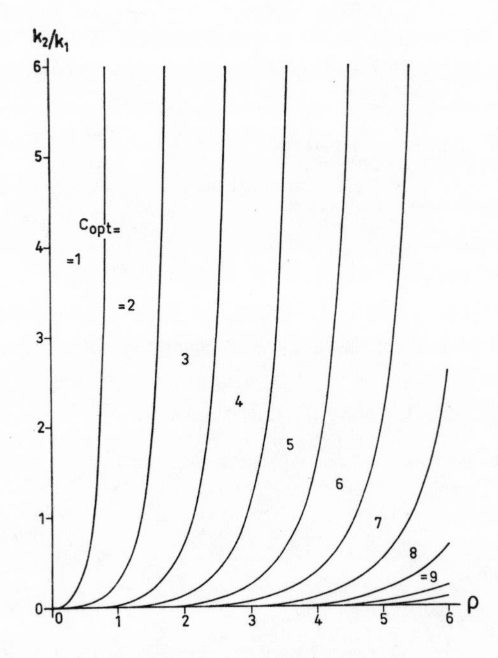

FIG. 2.5. Curves of indifference for the optimal number of servers in a delay system.

for various values of the parameter c $(1, 2, \ldots)$. The curves are called *curves of indifference*. They separate areas in which the optimal number of servers is $1, 2, \ldots$ So, when $\varrho = 4.2$ and $k_2/k_1 = 1.5$, the optimal number of servers is 6, as the condition (2.4.3) is satisfied for $c = 2, \ldots, 6$.

The results of this section are strongly connected with *Moe's principle* (cf. Arne Jensen, 1950).

3. THE DISTRIBUTION OF WAITING-TIMES IN THE SYSTEM *M/M/c*-DELAY

THE average waiting-time W of delayed demands does not depend on the order in which waiting demands are given service, as long as this order is not influenced by some advance knowledge of the holding-times of waiting demands. The distribution of the waiting-time $\underline{W}$, however, depends on this order. Let $q(\tau)$ and $Q(\tau)$ be the p.d.f. and c.d.f. of $\underline{W}$ for delayed demands. In the next three sections those quantities will be evaluated for the queue-disciplines "first-come-first-served", "last-come-first-served" and "random" (cf. Section 1.1.3).

3.1. First-come-first-served

When a demand arrives that finds all servers occupied, the number s of demands ahead of the new demand in the queue may have any value. First of all we shall evaluate $Q_s(\tau)$, the c.d.f. of the waiting-time $\underline{W}$ of a demand that finds s queued demands ahead of it. Now, in order that $\underline{W}$ for the arriving demand be less than τ, all s waiting demands plus the newly arriving one must have left the queue during τ. So this is equivalent to the ending of at least $s + 1$ occupations during τ. As the servers are fully occupied during this process, the ends of occupations form a Poisson process with a rate c (bearing in mind that the holding-time is exponential (1)). So, according to (1.3.5), changing ϱ into c, one obtains:

$$Q_s(\tau) = \sum_{i=s+1}^{\infty} \frac{e^{-c\tau}(c\tau)^i}{i!}, \tag{3.1.1}$$

The probability that a demand suffering delay finds s queued demands ahead of it can be evaluated with the aid of (2.3.4). It is

$$p_{c+s} \Big/ \sum_{s=0}^{\infty} p_{c+s} = (1-\eta)\,\eta^{s} \quad \text{with} \quad \eta = \varrho/c. \tag{3.1.2}$$

Hence the unconditional c.d.f. $Q(\tau)$ of the waiting-time $\underline{W}$ is:

$$Q(\tau) = \sum_{s=0}^{\infty} (1-\eta)\,\eta^{s} Q_{s}(\tau) = \sum_{s=0}^{\infty} (1-\eta)\,\eta^{s} \sum_{i=s+1}^{\infty} e^{-c\tau} \frac{(c\tau)^{i}}{i!}$$

$$= e^{-c\tau} \sum_{i=1}^{\infty} \frac{(c\tau)^{i}}{i!} \sum_{s=0}^{i-1} (1-\eta)\,\eta^{s} = e^{-c\tau} \sum_{i=0}^{\infty} \frac{(c\tau)^{i}}{i!} (1-\eta^{i}).$$

Thus

$$Q(\tau) = 1 - e^{-(c-\varrho)\tau}. \tag{3.1.3}$$

Under the condition "first-come-first-served" the waiting-time $\underline{W}$ is exponential with average $W = 1/(c-\varrho)$; the latter fact is in agreement with (2.3.9).

3.2. Last-come-first-served

A newly arriving demand that cannot be served immediately is observed; let it be denoted by A. At some instant at which A is still waiting, let s be the number of waiting demands that have arrived after A. Let the situation be described by saying that "the system is in [s]". When A arrives the system is in [0]. Let $q_s(\tau)$ be the p.d.f. for the *further* waiting-time $\underline{f}$ for A, when the system is in [s]. Obviously, $q_0(\tau)$ equals the p.d.f. $q(\tau)$ for the (total) waiting-time for A.

Consider the event "$\underline{f} = \tau + \Delta\tau(+d\tau)$, whilst the system is in [s]". During the first part $\Delta\tau$ of the further waiting-time three events are possible when $s > 0$ (neglecting higher-order effects):

(i) one arrival (probability $\varrho\Delta\tau$); new state $[s+1]$;
(ii) one termination (probability $c\Delta\tau$); new state $[s-1]$;
(iii) no change (probability $1 - \varrho\Delta\tau - c\Delta\tau$); new state [s].

In order for $\underline{f}$ to be $\tau + \Delta\tau(+d\tau)$ in total, those events should be followed by the event that the further waiting-time is $\tau(+d\tau)$, *taken conditional on the new state*. The corresponding conditional probabilities are $q_{s+1}(\tau)\,d\tau$, $q_{s-1}(\tau)\,d\tau$ and $q_s(\tau)\,d\tau$, respectively. Hence:

$$q_s(\tau + \Delta\tau)\,d\tau \approx \varrho\Delta\tau \cdot q_{s+1}(\tau)\,d\tau + c\Delta\tau \cdot q_{s-1}(\tau)\,d\tau$$
$$+ (1 - \varrho\Delta\tau - c\Delta\tau) \cdot q_s(\tau)\,d\tau$$

or, passing to the limit $\Delta\tau \to 0$,

$$\frac{dq_s}{d\tau} = \varrho q_{s+1} - (\varrho + c)\,q_s + cq_{s-1} \quad (s = 1, 2, \ldots). \tag{3.2.1}$$

For $s = 0$ the event (ii) (termination) would lead to a situation with $f = 0$. Hence, for $s = 0$ and $\tau > 0$ the last term in (3.2.1) is absent (i.e. $q_{-1} = 0$).

The initial conditions are as follows. When the system is in $[s]$, with $s > 0$, it is impossible for the further waiting-time to end within the next interval $d\tau$. When the system is in [0] this probability is $cd\tau$. Hence,†

$$q_s(0) = c\delta_s^0. \tag{3.2.2}$$

Let us introduce Laplace Transforms: $q_s(\tau) \risingdotseq \mathfrak{q}_s(z)$ (cf. Appendix). Then (3.2.1) and (3.2.2) yield

$$z\mathfrak{q}_s(z) - c\delta_s^0 = \varrho\mathfrak{q}_{s+1} - (\varrho + c)\,\mathfrak{q}_s + c\mathfrak{q}_{s-1}. \tag{3.2.3}$$

Now, introduce a generating function $\mathfrak{q}_s(z) \triangleq \mathfrak{Q}(y, z)$ (cf. Appendix). Then,

$$z\mathfrak{Q}(y, z) - c = \frac{\varrho\mathfrak{Q}(y, z) - \varrho\mathfrak{Q}(0, z)}{y} - (\varrho + c)\,\mathfrak{Q}(y, z) + cy\mathfrak{Q}(y, z)$$

or

$$\mathfrak{Q}(y, z) = \frac{cy - \varrho\mathfrak{Q}(0, z)}{-cy^2 + (\varrho + c + z)\,y - \varrho}. \tag{3.2.4}$$

Now, as $q_s(\tau) \geqq 0$ and $\int_0^\infty q_s(\tau)\,d\tau = 1$, one has for Re $z \geqq 0$:

$$|\mathfrak{q}_s(z)| = \left|\int_0^\infty e^{-z\tau} q_s(\tau)\,d\tau\right| \leqq 1.$$

† The symbol δ_s^0 is the Kronecker symbol: $\delta_0^0 = 1$; $\delta_1^0 = \delta_2^0 = \cdots = 0$.

Then for $|y| \leqq 1 - \varepsilon \quad (\varepsilon > 0)$

$$|\mathfrak{Q}(y, z)| = |\sum_s \mathfrak{q}_s(z)\, y^s| \leqq 1/\varepsilon.$$

So, for $\operatorname{Re} z \geqq 0$ the function $\mathfrak{Q}(y, z)$ should be bounded within the circle $|y| = 1$. Now, consider the denominator of (3.2.4). In the complex y-plane it has the zeros:

$$\left.\begin{matrix}\xi_1(z)\\ \xi_2(z)\end{matrix}\right\} = \frac{\varrho + c + z}{2c} \pm \sqrt{\left(\frac{\varrho + c + z}{2c}\right)^2 - \frac{\varrho}{c}}. \tag{3.2.5}$$

For $z \to 0$, those zeros tend to 1 and ϱ/c, respectively. Now, $\varrho < c$. Hence, for $\operatorname{Re} z = \delta$ (small and positive) the zero $\xi_2(z)$ is within the circle $|y| = 1$. In order for $\mathfrak{Q}(y, z)$ to remain bounded under those circumstances, the numerator of (3.2.4) should vanish also for $y = \xi_2(z)$:

$$c\xi_2(z) - \varrho\mathfrak{Q}(0, z) = 0. \tag{3.2.6}$$

This then yields the unknown function $\mathfrak{Q}(0, z)$. Hence,

$$q(\tau) = q_0(\tau) \risingdotseq \mathfrak{q}_0(z) = \mathfrak{Q}(0, z) = \frac{c}{\varrho}\xi_2(z)$$

or with (3.2.5),

$$q(\tau) \risingdotseq \frac{1}{\varrho}\left\{\frac{\varrho + c + z}{2} - \sqrt{\left(\frac{\varrho + c + z}{2}\right)^2 - \varrho c}\right\}. \tag{3.2.7}$$

From existing tables of Laplace Transforms it can be derived (Carslaw and Jaeger, 1953, p. 356, no. 49) that

$$q(\tau) = \sqrt{\frac{c}{\varrho}}\, \frac{e^{-(\varrho + c)\tau}}{\tau} I_1(2\sqrt{\varrho c}\,\tau) \tag{3.2.8}$$

which is the p.d.f. of waiting-time under the "last-come-first-served" condition. Here I_1 is the modified Bessel function of order one. Beyond (3.2.2) the given derivation to a certain extent is a mere tribute to standard mathematical usage. For on one the hand, it is perfectly practicable to integrate (3.2.1) with (3.2.2) numerically in a straightforward way, using computers. On the other hand, the calculation of $q(\tau)$ from (3.2.8) entails the computation of Bessel functions. Now computers mostly "reduce" the evaluation of Bessel-functions to the integration of a set of differential equations!

The c.d.f. of the waiting-time under the "last-come-first-served" condition is

$$Q(\tau) = \sqrt{\frac{c}{\varrho}} \int_0^\tau \frac{1}{u} e^{-(\varrho+c)u} I_1(2\sqrt{\varrho c}\, u) du. \tag{3.2.9}$$

The behaviour for $\tau \to \infty$ is of particular interest. Application of the usual methods to the Laplace transform of $Q(\tau)$—i.e. $1/z$ times the right-hand member of (3.2.7)—yields a poor asymptotic expression. A better result is obtained by writing

$$1 - Q(\tau) = \sqrt{c/\varrho} \int_\tau^\infty \dots \tag{3.2.9'}$$

using $I_1(x) \sim e^x/\sqrt{2\pi x}$ and performing a partial integration. The final result is:

$$1 - Q(\tau) \sim \eta^{-3/4}\alpha^{1/2}\left[\frac{e^{-\alpha c\tau}}{\sqrt{\pi\alpha c\tau}} - 2 + 2\Phi(\sqrt{2\alpha c\tau})\right] \tag{3.2.10}$$

where $\eta = \varrho/c$, $\alpha = (1 - \sqrt{\eta})^2$ whilst Φ stands for the normal distribution. The problem was first treated by Vaulot (1954). The asymptotic expansion appears in a slightly different form in Riordan (1962).

3.3. "Random" condition

Again, a newly arriving demand A, that cannot be served immediately, is observed. The situation in the queue at some instant after arrival of A will be described by the following complete set of "states of the system":

$$[s] \begin{cases} s = 0, 1, 2, \dots, & A \text{ is waiting, together with } s \text{ others,} \\ s = * & A \text{ is no longer waiting.} \end{cases}$$

Let $Q_s(\tau)$ be the c.d.f. of the further waiting-time f for A, under the condition that the system is in state $[s]$. Obviously $Q_*(\tau) = 1$ for $\tau > 0$.

Consider the event "$f \leqq \tau + \Delta\tau$, whilst the system is in $[s]$, $s \geqq 0$". During the first part $\Delta\tau$ of the further waiting-time three events are possible (apart from higher-order effects):

(i) one arrival (probability $\varrho\Delta\tau$); new state $[s+1]$;
(ii) a server becomes free (probability $c\Delta\tau$) and is seized by one of the other waiters (conditional probability $s/(s+1)$); new state $[s-1]$;
(iii) a server becomes free (probability $c\Delta\tau$) and is seized by A (conditional probability $1/(s+1)$); new state $[*]$;
(iv) no change (probability $1 - \varrho\Delta\tau - c\Delta\tau$); new state $[s]$.

In order for f to be $\leqq \tau + \Delta\tau$ in total, these events should be followed by the event that the further waiting-time is $\leqq \tau$, *taken conditional on the new state*. The corresponding conditional probabilities are $Q_{s+1}(\tau)$, $Q_{s-1}(\tau)$, 1 and $Q_s(\tau)$, respectively. Hence:

$$Q_s(\tau + \Delta\tau) \approx \varrho\Delta\tau Q_{s+1}(\tau) + c\Delta\tau \cdot \frac{s}{s+1} \cdot Q_{s-1}(\tau) + c\Delta\tau \cdot \frac{1}{s+1} \cdot 1 + (1 - \varrho\Delta\tau - c\Delta\tau) \cdot Q_s(\tau),$$

or, passing to the limit $\Delta\tau \to 0$:

$$\frac{dQ_s}{d\tau} = \varrho Q_{s+1} - (\varrho + c)\, Q_s + \frac{cs}{s+1} Q_{s-1} + \frac{c}{s+1} \quad (s = 0, 1, 2, \ldots). \tag{3.3.1}$$

As the remaining waiting-time is $\geqq 0$ when $s \geqq 0$, the initial conditions are:

$$Q_s(0) = 0 \quad (s = 0, 1, 2, \ldots). \tag{3.3.2}$$

The set of differential eqns. (3.3.1) together with (3.3.2) can readily be solved numerically with the aid of a computer.

The probabilities for A to find s other waiters on arrival again is the expression in (3.1.2), as this probability is not influenced by the way in which waiting demands are chosen for service. Hence, once the conditional c.d.f.'s $Q_s(\tau)$ have been obtained for some τ, the unconditional c.d.f. can be calculated from

$$Q(\tau) = (1 - \eta) \sum_{s=0}^{\infty} \eta^s Q_s(\tau). \tag{3.3.3}$$

Equation (3.3.1) was derived by Vaulot (1946) and Palm (1946, 1957). An analytical solution stems from Pollaczek (1946). Approximations are given in Pollaczek (1946) and Riordan (1953).

3.4. Comparison of the cases "first-come-first-served", "last-come-first-served" and "random"

The results of Sections 3.1, 3.2 and 3.3 are shown in Figs. 3.1 and 3.2 for the values $\varrho = 0.5$ and 0.9, respectively. Moreover, for the case "last-come-first-served" also the result of the asymptotic approximation (3.2.10) is shown by circles. It appears to be remarkably good. Instead of the c.d.f. $Q(\tau)$, its complement $1 - Q(\tau)$—i.e. the probability of a waiting-time *in excess* of τ—has been drawn.

In general it cannot be decided whether "first-come-first-served" or "last-come-first-served" queue-discipline is better. Mostly, however, it is important to guarantee that some rather large critical value of waiting-time is not exceeded too frequently. In this case the right-hand portions of the curves should be compared, and a "first-come-first-served" discipline is best.

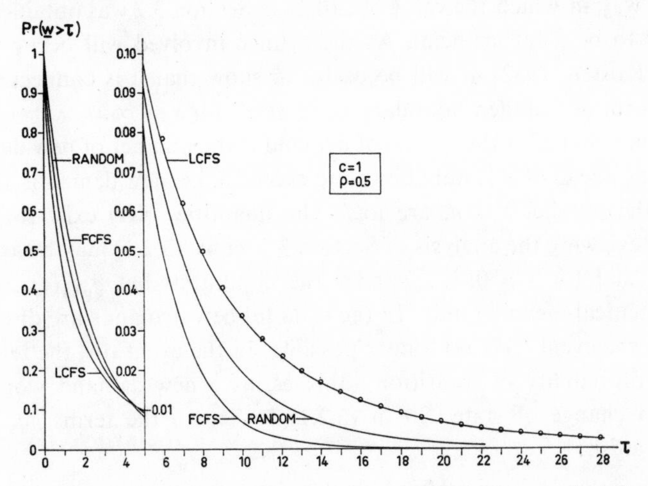

Figs. 3.1. Comparison of waiting-time distributions for the queue-disciplines "first-come-first-served", "last-come-fist-served" and "random".

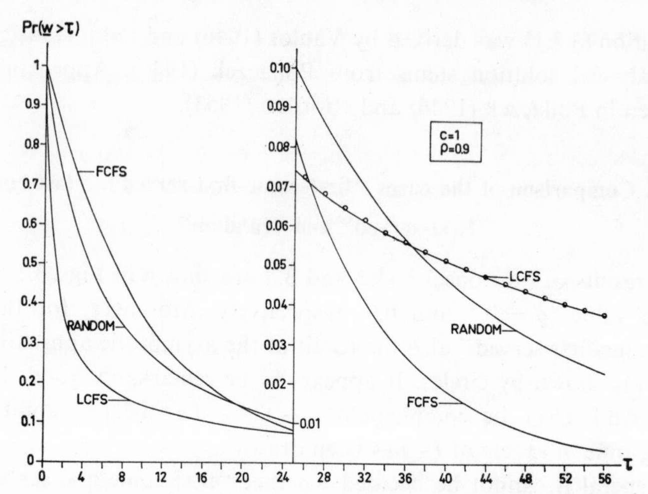

FIG. 3.2. Comparison of waiting-time distributions for the queue-disciplines "first-come-first-served", "last-come-first-served" and "random".

3.5. A note on analytical methods

The way in which the value of $\mathfrak{Q}(0, z)$ in Section 3.2 was obtained, may appear to be a bit artificial. As the artifice involved will occur several times (Kosten, 1952), it will be useful to show that it is connected with some form of "hidden boundary condition" for $s \to \infty$.

Assume that after the arrival of demand A the number of new demands queueing ahead of A is not allowed to exceed n, i.e. new demands arriving when the number $s = n$, are lost. The quantities $q_s(\tau)$ exist for $s \leqq n$ only. Reviewing the analysis of Section 3.2, eqns. (3.2.1) may be assumed to be valid for $s = 0, 1, \ldots$ under the condition that $q_{n+1}(\tau), \ldots$ have mathematical meaning only. In the state $[n]$ new demands are discarded. Hence, the event (i) is no longer possible. In the event (iii) the term $\varrho\Delta\tau$ in the probability of transition vanishes, as a new demand would not cause a change of state. So in (3.2.1) for $s = n$ the terms ϱq_{n+1} and $-\varrho q_n$ vanish:

$$\frac{dq_n}{d\tau} = -cq_n + cq_{n-1}. \tag{3.5.1}$$

Now we have assumed (3.2.1) to be valid too for $s = n$. Subtraction yields

$$q_{n+1}(\tau) = q_n(\tau). \tag{3.5.2}$$

The analysis up to (3.2.4) remains formally valid. The last result may be split into simple fractions [ξ_1 and ξ_2 are given by (3.2.5)]:

$$\mathfrak{Q}(y, z) = \frac{A(z)}{y - \xi_1} + \frac{B(z)}{y - \xi_2}, \tag{3.5.3}$$

where

$$B(z) = \frac{\xi_2 - \varrho\mathfrak{Q}(0, z)/c}{\xi_1 - \xi_2}. \tag{3.5.4}$$

From (3.5.3) it follows that

$$\mathfrak{q}_s(z) = -A(z)\,\xi_1^{-s-1} - B(z)\,\xi_2^{-s-1}. \tag{3.5.5}$$

Now, (3.5.2) yields $\mathfrak{q}_{n+1}(z) = \mathfrak{q}_n(z)$. Hence:

$$B = -\left(\frac{\xi_2}{\xi_1}\right)^{n+2} \cdot \frac{1 - \xi_1}{1 - \xi_2} A. \tag{3.5.6}$$

In the vicinity of $z = 0$ we have $|\xi_2/\xi_1| \approx \varrho/c < 1$. So for $n \to \infty$, B tends to zero. Then (3.5.4) shows that

$$\mathfrak{Q}(0, z) = \frac{c}{\varrho}\xi_2(z).$$

From here on the analysis of Section 3.2 may be resumed.

4. GENERAL HOLDING-TIME ASSUMPTION (*CM*-CLASS)

In the derivation of the birth-and-death equations for the "standard cases" in Chapter 2 the total termination rate for a state with r occupied servers has been taken to be r. This was done in virtue of the assumption of the holding-time to be exponential (1). If nothing were known about the ages of those r occupations, the total termination rate would still be r independent of the distribution of holding-times. In this case the results for the standard cases would be valid for any distribution of holding-times!

Now, there is a pitfall to be avoided. We cannot indicate clearly in what way the specifying of the number of occupied servers in a state would convey to us inferential knowledge as to the ages of those occupations. This, however, is no pretext for closing our eyes to the possibility of such a correlation and declaring the results of Chapter 2 to be "general" in consequence! Advocating this viewpoint is as silly as considering everybody guilty of murder, unless he proves his innocence!

In the next sections the general holding-time cases $M/G/c$-blocking and $M/G/1$-delay will be dealt with. The $M/G/c$-delay case has been treated by Pollaczek (1930, 1934, 1961). The theory uses very complicated analytical tools. A very good simplified explanatory version has been given by Syski (1967). This case will be given some attention in Section 4.3.

The way in which the problems of the next sections are tackled is by the so-called method of *supplementary variables* (used by the author since (1942, 1948a); the name is due to Cox (1955)). When the simple methods of Sections 2.1 and 2.2 do not work because of the fact that the termination rate is possibly dependent in some vague way on the past durations of the occupations, it is rational to subdivide the states by specifying the exact values of the past durations ("*markovization*"). When doing so, one can now use the conditional termination rates. This yields exact, though sometimes very complicated, relations.

In addition in Sections 4.4 and 4.5 the use of so-called *Erlang and hyperexponential distributions* will be discussed.

4.1. The case *M/G/c*-blocking

There are c identical servers, which can help the demands of a Poisson arrival process (arrival rate ϱ). The holding-time has a general distribution (*CM*-class) with an average 1. Stationarity is assumed.

Let us denote by $[r;\ \tau_1(+d\tau_1)\ \ldots\ \tau_r(+d\tau_r)]$ the state where r servers are occupied whilst the past durations of those r occupations are $\tau_1(+d\tau_1)$, ..., $\tau_r(+d\tau_r)$. The order of occupations is governed by the following prescription. Whenever a demand is made when r ($<c$) occupations $0_1, \ldots, 0_r$ (ranked in that order) are present, the new occupation O is given a place amongst the existing ones in random order. Hence, after the insertion of O the $r+1$ orders $(\mathrm{O}\,0_1, \ldots, 0_r)$, $(0_1\mathrm{O}0_2, \ldots, 0_r)$ ..., $(0_1, \ldots, 0_r\mathrm{O})$ are equiprobable. By this fictitious ordering mechanism the probability $q_r(\tau_1, \ldots, \tau_r)\, d\tau_1 \ldots d\tau_r$ of the state $[r;\ \tau_1(+d\tau_1) \ldots \tau_r(+d\tau_r)]$ may be supposed to be a symmetrical function of its arguments $\tau_1, \ldots, \tau_r$.

Now, consider the interval $(\tau, \tau + \Delta\tau)$ and suppose that at $\tau + \Delta\tau$ the system is in $[r; \tau_1 + \Delta\tau(+d\tau_1) \ldots \tau_r + \Delta\tau(+d\tau_r)]$, $(r < c)$; probability $q_r(\tau_1 + \Delta\tau, \ldots, \tau_r + \Delta\tau)\, d\tau_1 \ldots d\tau_r$. As the arguments are $\geqq \Delta\tau$, none of the r occupations has begun during $(\tau, \tau + \Delta\tau)$.

Hence there are the following possibilities:

(i) at time τ the state was $[r;\ \tau_1(+d\tau_1) \ldots \tau_r(+d\tau_r)]$, (probability $q_r(\tau_1, \ldots, \tau_r)\, d\tau_1 \ldots d\tau_r$); no new demand (probability $1 - \varrho\Delta\tau$); none of the r occupations has terminated (probability $\prod_{i=1}^{r} \{1 - g(\tau_i)\Delta\tau\}$);

(ii) at time τ the state was $[r+1;\ \tau_1(+d\tau_1) \ldots \tau_r(+d\tau)\ u(+du)]$ (probability $q_{r+1}(\tau_1, \ldots, \tau_r, u)\, d\tau_1 \ldots d\tau_r du$); the extra duration u may have any value (integration); moreover, the extra occupation may have had any of $r+1$ equiprobable positions (factor $r+1$).

The possible states, transitions and probabilities may be summarized as follows:

[State] and probability at time τ	Transition and probability
$\left\{\begin{matrix}[r;\ \tau_1(+d\tau_1) \ldots \tau_r(+d\tau_r)] \\ q_r(\tau_1, \ldots, \tau_r)\, d\tau_1 \ldots d\tau_r\end{matrix}\right\}$	$\left\{\begin{matrix}\text{no demand; no termination} \\ (1-\varrho\Delta\tau)\cdot\prod_{i=1}^{r}\{1-g(\tau_i)\Delta\tau\}\end{matrix}\right\}$
$\left\{\begin{matrix}[r+1;\ \tau_1(+d\tau_1) \ldots \tau_r(+d\tau_r)\ u(+du)] \\ q_{r+1}(\tau_1, \ldots, \tau_r, u)\, d\tau_1 \ldots d\tau_r du \\ \text{(all } u,\ r+1 \text{ transpositions)}\end{matrix}\right\}$	$\left\{\begin{matrix}\text{termination extra occupation} \\ g(u)\,\Delta\tau\end{matrix}\right\}$ $\Rightarrow$

[State] and probability at time $\tau + \Delta\tau$

$$\Rightarrow \left\{\begin{matrix}[r;\ \tau_1+\Delta\tau(+d\tau_1) \ldots \tau_r+\Delta\tau(+d\tau_r)] \\ q_r(\tau_1+\Delta\tau, \ldots, \tau_r+\Delta\tau)\, d\tau_1 \ldots d\tau_r\end{matrix}\right\}.$$

This leads to:

$$q_r(\tau_1+\Delta\tau, \ldots, \tau_r+\Delta\tau)\, d\tau_1 \ldots d\tau_r \approx q_r(\tau_1, \ldots, \tau_r)\, d\tau_1 \ldots d\tau_r \cdot (1-\varrho\Delta\tau)\times$$
$$\times \prod_{i=1}^{r}\{1-g(\tau_i)\Delta\tau\} + (r+1)\int_{u=0}^{\infty} q_{r+1}(\tau_1, \ldots, \tau_r, u)\, d\tau_1 \ldots d\tau_r du\, g(u)\,\Delta\tau.$$

The left-hand side member is developed in powers of $\Delta\tau$ and the limit $\Delta\tau \to 0$ is taken. This yields

$$\sum_{i=1}^{r} \frac{\partial}{\partial\tau_i} q_r(\tau_1, \ldots, \tau_r) = -\left\{\varrho + \sum_{i=1}^{r} g(\tau_i)\right\} q_r(\tau_1, \ldots, \tau_r)$$
$$+ (r+1)\int_0^{\infty} q_{r+1}(\tau_1, \ldots, \tau_r, u)\, g(u)\, du \quad (r = 0, \ldots, c-1). \qquad (4.1.1)$$

Now assume that the state at time $\tau + \Delta\tau$ is $[r+1; \tau_1 + \Delta\tau(+d\tau_1) \ldots \tau_r + \Delta\tau(+d\tau_r), 0(+\Delta\tau)]$ (probability $q_{r+1}(\tau_1 + \Delta\tau, \ldots, \tau_r + \Delta\tau, 0)\, d\tau_1 \ldots d\tau_r \Delta\tau$). Then the state at time τ was $[r; \tau_1(+d\tau_1) \ldots \tau_r(+d\tau_r)]$ (probability $q_r(\tau_1, \ldots, \tau_r)\, d\tau_1 \ldots d\tau_r$). The transition then must consist of (i) a new demand (probability $\varrho\Delta\tau$), and of (ii) this demand leading to an occupation in the proper one of $r+1$ possible positions (probability $1/(r+1)$). Hence,

$$q_{r+1}(\tau_1 + \Delta\tau, \ldots, \tau_r + \Delta\tau, 0)\, d\tau_1 \ldots d\tau_r \Delta\tau$$
$$= q_r(\tau_1, \ldots, \tau_r)\, d\tau_1 \ldots d\tau_r \cdot \varrho\Delta\tau \cdot \frac{1}{r+1} + O(\Delta\tau^2)$$

or when $\Delta\tau \to 0$,

$$(r+1)\, q_{r+1}(\tau_1, \ldots, \tau_r, 0) = \varrho q_r(\tau_1, \ldots, \tau_r) \quad (r = 0, \ldots, c-1). \quad (4.1.2)$$

Finally the introduced states form a complete and mutually exclusive set:

$$\sum_{r=0}^{c} \int_0^\infty \cdots_{(r)} \int_0^\infty q_r(\tau_1, \ldots, \tau_r)\, d\tau_1 \ldots d\tau_r = 1. \quad (4.1.3)$$

Let us assume that the derived equations possess a unique solution. Then try the method of separation of variables. In view of the symmetry of the q_r functions one can try:

$$q_r(\tau_1, \ldots, \tau_r) = C_r \Psi(\tau_1) \ldots \Psi(\tau_r) \quad (4.1.4)$$

where Ψ is an unknown function and C_r an unknown constant. We can put $\Psi(0) \equiv 1$ without loss of generality. Substitution into (4.1.2) yields

$$(r+1)\, C_{r+1} = \varrho C_r \quad (r = 0, \ldots, c-1). \quad (4.1.5)$$

Multiple application shows that

$$C_r = \frac{\varrho^r}{r!} C_0 \quad (4.1.6)$$

and hence

$$q_r(\tau_1, \ldots, \tau_r) = C_0 \frac{\varrho^r}{r!} \Psi(\tau_1) \ldots \Psi(\tau_r). \quad (4.1.7)$$

Insertion into (4.1.1) and some rearrangement yields:

$$\sum_{i=1}^{r}\left\{\frac{d}{d\tau_i}\ln\Psi(\tau_i)+g(\tau_i)\right\}=\varrho\left\{-1+\int_0^\infty \Psi(u)\,g(u)\,du\right\}\;(r=0,\ldots,c-1). \tag{4.1.8}$$

This is only possible when both the right-hand member and all terms of the sum vanish:

$$\int_0^\infty \Psi(u)\,g(u)\,du = 1, \tag{4.1.9}$$

$$\frac{d}{d\tau}\ln\Psi(\tau) = -g(\tau). \tag{4.1.10}$$

With the aid of (1.2.14) and (1.2.17):

$$\Psi(\tau) = K\int_\tau^\infty f(u)\,du. \tag{4.1.11}$$

As $\Psi(0) = 1$, K must be 1. Then the left-hand member of (4.1.9) is equal to $\int_0^\infty f(u)\,du$ (cf. 1.2.15). So (4.1.9) is an identity.

Insertion of (4.1.11) into (4.1.7) yields

$$q_r(\tau_1,\ldots,\tau_r) = C_0\frac{\varrho^r}{r!}\prod_{i=1}^{r}\int_{\tau_i}^\infty f(u_i)\,du_i. \tag{4.1.12}$$

The probability p_r of the "overall state" $[r]$ (i.e. without stipulation of the past durations) is obtained by integration:

$$p_r = C_0\frac{\varrho^r}{r!}\prod_{i=1}^{r}\int_0^\infty\left(\int_{\tau_i}^\infty f(u)\,du\right)d\tau_i.$$

The double integral equals

$$\int_0^\infty dv\int_v^\infty f(u)\,du = v\int_v^\infty f(u)\,du\,\Big|_{v=0}^{\infty} + \int_0^\infty vf(v)\,dv = 0+1 = 1.$$

Hence,

$$p_r = C_0 \varrho^r / r!. \tag{4.1.13}$$

Now, (4.1.3) is equivalent to $\sum_0^c p_r = 1$. Hence,

$$1/C_0 = \sum_{r=0}^{c} \varrho^r / r!. \tag{4.1.14}$$

This then yields the results (2.1.8) and (2.1.9) for p_r and B, respectively. Hence, *the Erlang B-formula is valid under general holding-time assumption.* The original proof is by Palm (1938). The proof given above is a modification of the proof by Kosten (1948). Another demonstration stems from Sevastyanov (1957). Erlang's original paper (1918) contains a proof for the case of constant holding-time.

4.2. The case $M/G/1$-delay

The suspicion might rise that the independence of the Erlang B-formula of the distribution of holding-time is in some way trivial. But then it is to be expected that the results of Section 2.3 are also independent of this distribution. This, however, is not the case, as will result from the following treatment of the $M/G/1$-delay case.

The following complete set of states and their stationary probabilities is introduced:

State	Description	Probability
$[-]$	server idle; queue empty	p
$[r, \tau, d\tau]$	server busy; past duration $\tau(+d\tau)$; $\underline{r} = r$ in queue ($\geqq 0$)	$q_r(\tau)\, d\tau$

Consider the interval $(\tau', \tau' + \Delta\tau)$ and suppose the system to be in $[-]$ at $\tau' + \Delta\tau$. Then at time τ' the state was either $[-]$ followed by "no demand" or $[0, u, du]$ (u arbitrary) followed by "termination of occupation". This leads to the following transition scheme for the end state [−]:

[State] and probability at time τ'	Transition and probability	[State] and probability at time $\tau' + \Delta\tau$
[−]	no demand →	
p	$1 - \varrho\Delta\tau$	[−]
[0, u, du] (all u)	termination →	p
$\int q_0(u) \ldots du$	$g(u)\,\Delta\tau$	

from which

$$p = p(1 - \varrho\Delta\tau) + \int_0^\infty q_0(u)\,g(u)\,\Delta\tau du + O(\Delta\tau^2)$$

or

$$\varrho p = \int_0^\infty q_0(u)\,g(u)\,du. \tag{4.2.1}$$

The transition scheme for the end state $[r, \tau + \Delta\tau, d\tau]$ is:

[State] and probability at time τ'	Transition and probability ↓	[State] and probability at time $\tau' + \Delta\tau$
$[r, \tau, d\tau]$	no demand; no termination →	
$q_r(\tau)\,d\tau$	$(1 - \varrho\Delta\tau) \cdot \{1 - g(\tau)\,\Delta\tau\}$	$[r, \tau + \Delta\tau, d\tau]$
$[r - 1, \tau, d\tau]$	new demand →	$q_r(\tau + \Delta\tau)\,d\tau$
$q_{r-1}(\tau)\,d\tau$	$\varrho\Delta\tau$	

which yields

$$q_r(\tau + \Delta\tau)\,d\tau = q_r(\tau)\,d\tau \cdot (1 - \varrho\Delta\tau) \cdot \{1 - g(\tau)\,\Delta\tau\} + q_{r-1}(\tau)\,d\tau \cdot \varrho\Delta\tau + O(\Delta\tau^2)$$

or, in the limit,

$$\frac{dq_r(\tau)}{d\tau} = -\{\varrho + g(\tau)\}\, q_r(\tau) + \varrho q_{r-1}(\tau). \tag{4.2.2}$$

Now, consider the state $[r, 0, \Delta\tau]$ at time $\tau' + \Delta\tau$. At time τ' the present occupation did not exist. Hence, the queue can have had a length $r + 1$, the former occupation (any past duration u) has ended during $\Delta\tau$ and the present occupation taken its place. When, however, $r = 0$, it is also possible that at time τ' the state was $[-]$ and that the present occupation stems from a newcomer:

[State] and probability at time τ'	Transition and probability	[State] and probability at time $\tau' + \Delta\tau$
$[r+1, u, du]$ (all u)	termination $\longrightarrow$	
$\int q_{r+1}(u) \dots du$	$g(u)\,\Delta\tau$	$[r, 0, \Delta\tau]$
$[-]$	$(r = 0)$ demand $\longrightarrow$	$q_r(0)\,\Delta\tau$
p	$\delta_r^0 \cdot \varrho\Delta\tau$	

The result is

$$q_r(0) = \int_0^\infty q_{r+1}(u)\, g(u)\, du + \varrho p \delta_r^0. \tag{4.2.3}$$

Finally, the states form a complete set of disjoint events:

$$p + \sum_{r=0}^{\infty} \int_0^\infty q_r(u)\, du = 1. \tag{4.2.4}$$

We now replace the functions $q_r(\tau)$ by their generating function $Q(x, \tau)$. With the aid of (A 3, 4, 8 and 9) the eqns. (4.2.1–4) transform into

$$\varrho p = \int_0^\infty Q(0, u)\, g(u)\, du, \tag{4.2.5}$$

$$\frac{\partial Q}{\partial \tau} = \{\varrho x - \varrho - g(\tau)\}\, Q(x, \tau), \tag{4.2.6}$$

$$Q(x, 0) = \frac{1}{x} \int_0^\infty \{Q(x, u) - Q(0, u)\}\, g(u)\, du + \varrho p, \tag{4.2.7}$$

$$p + \int_0^\infty Q(1, u)\, du = 1. \tag{4.2.8}$$

With the notation (1.2.14) the general solution of (4.2.6) may be written as

$$Q(x, \tau) = K(x) \exp\{-\varrho(1 - x)\,\tau - G(\tau)\}. \tag{4.2.9}$$

Using this expression, with the relations (1.2.15) and (1.2.17), and introducing the notation

$$J(x) := \int_0^\infty e^{-\varrho(1-x)u} f(u)\, du, \tag{4.2.10}$$

one obtains

$$\int_0^\infty Q(x, u)\, g(u)\, du = K(x)\, J(x). \tag{4.2.11}$$

From (4.2.10)

$$J(1) = 1. \tag{4.2.12}$$

Eliminating p between (4.2.5) and (4.2.7) and then using (4.2.9) and (4.2.11) one obtains ($e^{-G(0)} = 1$)

$$K(x) = \frac{1}{x}\{K(x)\, J(x) - K(0)\, J(0)\} + K(0)\, J(0),$$

and hence:

$$K(x) = \frac{(1 - x)\, K(0)\, J(0)}{J(x) - x}. \tag{4.2.13}$$

Inserting this expression into (4.2.9) and comparing (4.2.11) and (4.2.5) yields both unknowns up to a constant $K(0)$:

$$Q(x, \tau) = \frac{(1 - x)\, K(0)\, J(0)}{J(x) - x} \cdot \exp\{-\varrho(1 - x)\tau - G(\tau)\}, \quad (4.2.14)$$

$$p = \frac{1}{\varrho} K(0)\, J(0). \quad (4.2.15)$$

Those expressions are inserted into the norming eqn. (4.2.8), which yields

$$K(0)\, J(0) = 1 \Big/ \left[\frac{1}{\varrho} + \lim_{x \to 1} \frac{1 - x}{J(x) - x}\right].$$

The limit is obtained by l'Hôpital's rule:

$$K(0)\, J(0) = 1 \Big/ \left[\frac{1}{\varrho} - \frac{1}{J'(1) - 1}\right]. \quad (4.2.16)$$

Now (cf. 4.2.10),

$$J'(1) = \varrho \int_0^\infty u f(u)\, du = \varrho, \quad (4.2.17)$$

and hence,

$$K(0)\, J(0) = \varrho(1 - \varrho). \quad (4.2.18)$$

Thus all probabilities of states are known. From (4.2.15) and (4.2.18) the probability of delay D ensues:

$$D = 1 - p = \varrho. \quad (4.2.19)$$

This result is obvious. Per unit of time the server is engaged on the average for $\varrho \cdot 1 = \varrho$ units of time. Hence, the probability of delay for a virtual demand is ϱ, irrespective of the distribution of holding-time.

The average waiting-time is calculated as follows. The expectation of total waiting-time $E(\underline{w}_{tot})$, incurred by all queued demands per unit of time, is (cf. A 6 and 9):

$$E(\underline{w}_{tot}) = E(\underline{r}) \cdot 1 = \sum_{r=0}^{\infty} r \int_0^\infty q_r(u)\, du = x \frac{d}{dx} \int_0^\infty Q(x, u)\, du \big|_{x=1}.$$

Insertion of (4.2.14), (4.2.10) and (4.2.18) yields

$$E(\underline{w}_{\text{tot}}) = \varrho(1-\varrho)\frac{d}{dx}\left.\frac{(1-x)\int_0^\infty \exp\{-\varrho(1-x)u - G(u)\}\,du}{\int_0^\infty \exp\{-\varrho(1-x)u\}f(u)\,du - x}\right|_{x=1}. \tag{4.2.20}$$

After some tedious analysis, with the use of l'Hôpital's rule and the result that

$$\int_0^\infty ue^{-G(u)}du = \frac{u^2}{2}e^{-G(u)}\Big|_0^\infty + \frac{1}{2}\int_0^\infty u^2 f(u)\,du = \frac{1}{2}E(\underline{h}^2),$$

one obtains

$$E(\underline{w}_{\text{tot}}) = \frac{\varrho^2}{2(1-\varrho)}E(\underline{h}^2). \tag{4.2.21}$$

The average number of demands per unit of time is ϱ, the average number of delayed demands $\varrho D = \varrho^2$. When $E(\underline{w}_{\text{tot}})$ is divided by those quantities we obtain the average waiting time per demand (w) and per delayed demand (W), respectively,

$$w = \frac{\varrho}{2(1-\varrho)}E(\underline{h}^2),$$
$$W = \frac{1}{2(1-\varrho)}E(\underline{h}^2). \tag{4.2.22}$$

When times are not normed, these expressions must be divided by h. These well-known results stem from Pollaczek (1930); sometimes they are attributed to Khintchine (1932).

4.3. A note on the system $M/G/c$-delay

Let the arrival process be Poissonian with arrival rate ϱ, whilst the number of servers is c. The holding-time has an arbitrary distribution (CM-class) with average 1. The following complete set of states and their probabilities is introduced:

State	Description	Probability
$[r; \tau_1(+d\tau_1) \ldots \tau_r(+d\tau_r)]$ $(0 \leq r < c)$	r servers engaged; the holding-times are $\tau_1(+d\tau_1), \ldots, \tau_r(+d\tau_r)$	$q_r(\tau_1, \ldots, \tau_r)\, d\tau_1 \ldots d\tau_r$,
$[c + s; \tau_1(+d\tau_1) \ldots \tau_c(+d\tau_c)]$ $(s = 0, 1, \ldots)$	all c servers engaged; s demands in queue; the holding-times are $\tau_1(+d\tau_1), \ldots, \tau_c(+d\tau_c)$	$q_{c+s}(\tau_1, \ldots, \tau_c)\, d\tau_1 \ldots d\tau_c$.

Just as in Section 4.1, new occupations are supposed to take arbitrary places amongst the existing occupations, thus rendering q_r and q_{c+s} symmetrical functions of their arguments τ_i.

Much along the same lines as in the former sections the following set of equations of state may be obtained:

$$\sum_{i=1}^{r} \frac{\partial}{\partial \tau_i} q_r(\tau_1, \ldots, \tau_r) = -\left\{\varrho + \sum_{i=1}^{r} g(\tau_i)\right\} q_r(\tau_1, \ldots, \tau_r) + (r+1) \int_0^\infty q_{r+1}(\tau_1, \ldots, \tau_r, u)\, g(u)\, du, \quad (r = 0, \ldots, c-1) \tag{4.3.1}$$

$$\sum_{i=1}^{c} \frac{\partial}{\partial \tau_i} q_{c+s}(\tau_1, \ldots, \tau_c) = -\left\{\varrho + \sum_{i=1}^{c} g(\tau_i)\right\} q_{c+s}(\tau_1, \ldots, \tau_c) + \varrho q_{c+s-1}(\tau_1, \ldots, \tau_c) \quad (s = 0, 1, \ldots) \tag{4.3.2}$$

(delete last term for $s = 0$),

$$q_r(\tau_1, \ldots, \tau_{r-1}, 0) = \frac{\varrho}{r} q_{r-1}(\tau_1, \ldots, \tau_{r-1}) \quad (r = 1, \ldots, c-1), \tag{4.3.3}$$

$$q_{c+s}(\tau_1, \ldots, \tau_{c-1}, 0) = \int_0^\infty q_{c+s+1}(\tau_1, \ldots, \tau_{c-1}, u)\, g(u)\, du + \delta_s^0 \frac{\varrho}{c} q_{c-1}(\tau_1, \ldots, \tau_{c-1}) \quad (s = 0, 1, \ldots), \tag{4.3.4}$$

$$\sum_{r=1}^{\infty} \int_0^\infty \underset{(m)}{\cdots} \int_0^\infty q_r(\tau_1, \ldots, \tau_m)\, d\tau_1 \ldots d\tau_m = 1, \tag{4.3.5}$$

with $m = \text{Min}\,(r, c)$.

If the set of eqns. (4.3.1–5) could be solved, the probability of delay and the average waiting-time might be computed from

$$p_{c+s} := \int_0^\infty \underset{(c)}{\cdots} \int_0^\infty q_{c+s}(\tau_1, \ldots, \tau_c)\, d\tau_1 \ldots d\tau_c, \tag{4.3.6}$$

$$D = \sum_{s=0}^{\infty} p_{c+s}; \quad W = \frac{1}{\varrho D} \sum_{s=1}^{\infty} s p_{c+s}. \tag{4.3.7}$$

Cox (1955) states those equations for a slightly more general case. He also indicates a formal solution under certain assumptions.

The way in which Pollaczek tackles the $M/G/c$-delay case is entirely different. For the $M/D/c$-delay case cf. also Crommelin (1932).

4.4. Use of Erlang-*k* distributions

Suppose that the service rendered by a server consists of two activities, to be performed by the server *consecutively*. Assume that both activities have holding-times that possess a distribution which is exponential with an average $\frac{1}{2}$. Moreover, assume that the holding-times for both activities are mutually independent. Then the p.d.f. of the total holding-time $f(\tau)$ is equal to the convolution integral of two negative exponentials (for the definition of the "asterisk" product cf. A 27):

$$f(\tau) = (2e^{-2\tau}) * (2e^{-2\tau}) := \int_0^\tau (2e^{-2\tau_1})(2e^{-2(\tau-\tau_1)})\, d\tau_1 = 4\tau e^{-2\tau}.$$

The average can be verified to be 1 $(= \frac{1}{2} + \frac{1}{2})$.

Now, let us go further and assume the total activity to consist of k consecutive elementary activities all with mutually independent holding-times that are exponential $(1/k)$. The overall holding-time possesses a p.d.f. equal to a $(k-1)$-fold convolution of an exponential $(1/k)$ distribution (cf. A 28):

$$f(\tau) = (ke^{-k\tau})^{k*} := \int_0^\tau (ke^{-k\tau_1})\, d\tau_1 \int_0^{\tau-\tau_1} (ke^{-k\tau_2})\, d\tau_2 \ldots$$

$$\ldots \int_0^{\tau-\tau_1-\cdots-\tau_{k-2}} (ke^{-k\tau_{k-1}})(ke^{-k(\tau-\tau_1-\cdots-\tau_{k-1})})\, d\tau_{k-1}$$

$$= \cdots = \frac{k^k \tau^{k-1}}{(k-1)!} e^{-k\tau}, \tag{4.4.1}$$

which is called the *Erlang-k distribution*. The average is k times $1/k$, i.e. 1. It is identical to a *Chi-square distribution* with an even number of degrees of freedom (Saaty, 1961, p. 65).

In Fig. 4.1 a number of those distributions are shown. For $k = 1$ we have the well-known exponential (1) distribution. For $k \to \infty$ the distribution tends to constant (1).

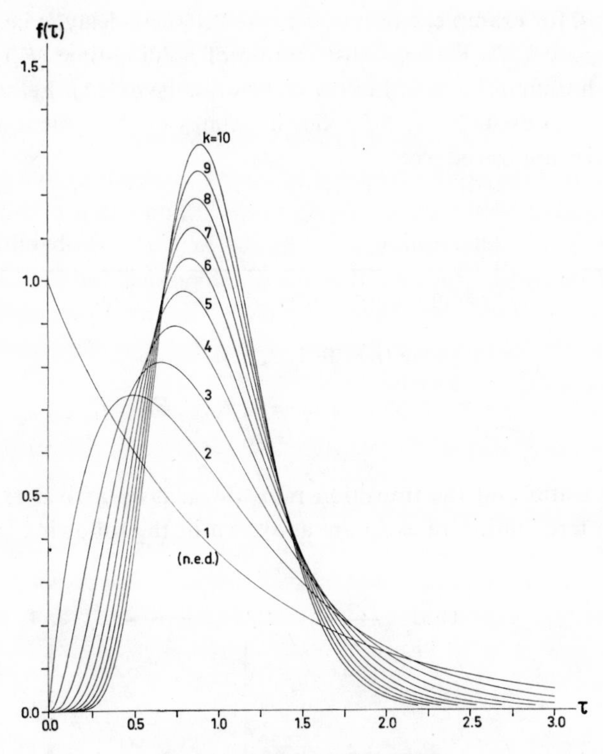

FIG. 4.1. Erlang-k distributions.

In introducing those distributions we have supposed that there are k subactivities with holding-times that are exponential ($1/k$). In cases where an Erlang-k distribution for the holding-time applies it is mostly not possible to point out k such subactivities. One can, however, think of k hypothetical phases 1, ..., k which the server assumes consecutively during one occupation. In any of those phases there is assumed a constant termination rate k. According to Section 1.2 the holding-times of those phases are then exponential ($1/k$). Each time a phase ends, the next phase is entered. When phase k ends, the occupation is at an end.

This (hypothetical) division in k phases for the Erlang-k distribution offers an alternative to deal with other distributions than the exponential one. Let us, for example, consider the case "$M/E_k/1$-delay"; i.e. Poisson arrival process (M), Erlang-k distribution of holding-time (E_k) and one server with (infinite) queue. Let the arrival rate be $\varrho\,(<1)$. Let us assume stationarity to exist. Now, the following states of the system and their probabilities are introduced:

State	Description	Probability
[−]	system empty	p
$[r, s]$	r in queue ($r = 0, 1, 2, \ldots$); server in phase s ($s = 1, 2, \ldots, k$)	f_{rs}

Then the states and the transition rates (formed by arrival rate ϱ and the phase termination rates k) are as shown in the following transition scheme:

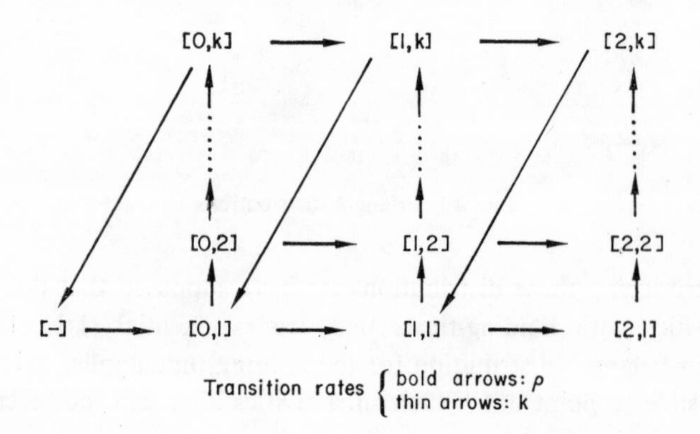

The equations of state are

$$0 = -\varrho p + kf_{0k}, \tag{4.4.2}$$

$$0 = -(\varrho + k)f_{r1} + \varrho f_{r-1,1} + kf_{r+1,k} + \delta_r^0 \varrho p, \tag{4.4.3}$$

$$0 = -(\varrho + k)f_{rs} + \varrho f_{r-1,s} + kf_{r,s-1}, \tag{4.4.4}$$

$$(r = 0, 1, \ldots; s = 2, \ldots, k; f_{-1,s} \equiv 0).$$

Let $f_{rs} \triangleq F_s(x)$. Then the last three equations yield

$$0 = -\varrho p + kF_k(0), \tag{4.4.5}$$

$$0 = -(\varrho - \varrho x + k)\, F_1(x) + k\, \frac{F_k(x) - F_k(0)}{x} + \varrho p, \tag{4.4.6}$$

$$0 = -(\varrho - \varrho x + k)\, F_s(x) + kF_{s-1}(x), \tag{4.4.7}$$

$$(s = 2, \ldots, k).$$

Multiple application of the recursion formula (4.4.7) yields

$$F_s(x) = \left\{1 + \frac{\varrho}{k}(1 - x)\right\}^{k-s} F_k(x) \quad (s = 1, \ldots, k). \tag{4.4.8}$$

Insertion of this expression for $F_1(x)$ in (4.4.6), and using the result $kF_k(0) = \varrho p$ [cf. (4.4.5)] yields

$$F_k(x) = \frac{\varrho p}{k}\, \frac{1 - x}{1 - x\{1 + \varrho(1 - x)/k\}^k}, \tag{4.4.9}$$

and hence

$$F_s(x) = \frac{\varrho p}{k}\, \frac{(1 - x)\,\{1 + \varrho(1 - x)/k\}^{k-s}}{1 - x\{1 + \varrho(1 - x)/k\}^k}. \tag{4.4.10}$$

Now, the defined states form a complete and mutually exclusive set. Hence,

$$p + \sum_{s=1}^{k} \sum_{r=0}^{\infty} f_{rs} = 1, \tag{4.4.11}$$

or

$$p + \sum_{s=1}^{k} F_s(1) = 1. \tag{4.4.12}$$

The relations (4.4.12) and (4.4.10) completely determine p and $F_s(x)$; the coefficients of the power-series expansion of the latter function yield the f_{rs}. This then constitutes the solution for the steady-state probabilities. From (4.4.10) $F_s(1)$ may be computed by l'Hôpital's rule, which yields $F_s(1) = \varrho p/k(1 - \varrho)$. Insertion in (4.4.12) yields

$$p = 1 - \varrho \tag{4.4.13}$$

which is the correct value (cf. Section 4.2).

The total waiting-time per unit of time is

$$E(\underline{w}_{tot}) = E(\underline{r}) \cdot 1 = \sum_{s=1}^{k} r f_{rs} = \sum_{s=1}^{k} \frac{d}{dx} F_s(x) \bigg|_{x=1} . \tag{4.4.14}$$

Use of (4.4.10) together with $p = 1 - \varrho$ yields (after some tedious algebra):

$$E(\underline{w}_{tot}) = \frac{\varrho^2}{1 - \varrho} \cdot \frac{k + 1}{2k}. \tag{4.4.15}$$

As the average number of delayed demands per unit of time is ϱ^2, the average waiting-time W per delayed demand is

$$W = E(\underline{w}_{tot})/\varrho^2 = \frac{k + 1}{2k} \cdot \frac{1}{1 - \varrho} . \tag{4.4.16}$$

Now,

$$E(\underline{h}^2) = \int_0^{\infty} \tau^2 \cdot \frac{k^k \tau^{k-1} e^{-k\tau}}{(k - 1)!} d\tau = \frac{k + 1}{k} . \tag{4.4.17}$$

Hence (4.4.16) is in concordance with the general result (4.2.22). It gives the average waiting-time for delayed demands in the case of a family of holding-time distributions, ranging from exponential (1) (viz. for $k = 1$) to constant (1) (for $k = \infty$). The variance of this distribution—i.e. $E(\underline{h}^2) - E^2(\underline{h}) = 1/k$—at the same time varies from 1 to 0.

When the method of supplementary variables is used (cf. Sections 4.1, 4.2 and 4.3) the system is rendered Markovian by those variables. This means that so much of the system's history is entered into the description of the present state, that the system's behaviour in the near future can be predicted in a statistical sense. For arbitrary holding-time distribution this necessitates entering a continuous argument per server (past duration of occupation). The method of the Erlang-k distributions is "cheaper" in this respect. It needs the specification of the phase of the server $(1, \ldots, k)$ only (discrete argument). Whereas the method of the supplementary variables is able to cope with *any* p.d.f., the Erlang-k method leaves a choice from a family of p.d.f.'s which have two characteristics:

(i) they have one maximum;
(ii) for $k > 1$ their variance is *essentially less* than for the exponential (1) case.

In many instances these restrictions are not serious. Before using the method it should, however, be investigated whether statistical material concerning holding-times allows the assumption of the Erlang-k distribution. The method used in this section conforms to Morse (1958).

4.5. Hyperexponential distributions

Suppose the flow of demands stems from two groups of sources. Both groups lead to exponential holding-times, with different averages, however. Let a fraction σ of the demands lead to type A occupations, the holding-time of which is exponential (h_1), whereas the remaining fraction $1 - \sigma$ leads to type B occupations with exponential (h_2) holding-time. The overall average holding-time then is equal to $\sigma h_1 + (1 - \sigma) h_2$. When we take $h_1 = 1/2\sigma$ and $h_2 = 1/2(1 - \sigma)$, this average is unity. The p.d.f. in this case is

$$f(\tau) = \sigma \cdot \frac{1}{h_1} e^{-\tau/h_1} + (1 - \sigma) \cdot \frac{1}{h_2} e^{-\tau/h_2}$$

$$= 2\sigma^2 e^{-2\sigma\tau} + 2(1 - \sigma)^2 e^{-2(1-\sigma)\tau}, \qquad (4.5.1)$$

a so-called *hyperexponential distribution* Morse, 1958). When $\sigma \gg 1 - \sigma$ (and hence $h_1 \ll h_2$) the shape of the p.d.f. is mainly determined by the exponential (h_1) component, but for the "tail" that reveals the exponential (h_2) component.

When $\sigma = \frac{1}{2}$, both exponentials are identical and $f(\tau)$ reduces to the exponential (1) case. In all other cases, however, the variance is larger than in the exponential (1) case (i.e. >1):

$$\operatorname{var}(\underline{h}) = \int_0^\infty 2\tau^2\{\sigma^2 e^{-2\sigma\tau} + (1-\sigma)^2 e^{-2(1-\sigma)\tau}\}\, d\tau - 1$$

$$= \tfrac{1}{2}\left(\frac{1}{\sigma} + \frac{1}{1-\sigma}\right) - 1. \qquad (4.5.2)$$

It is possible that the shape of a p.d.f. resembles the hyperexponential distribution (4.5.1), though one may not be able to detect two occupation types with different exponential holding-times. In this case one may hypothetically split up the flow of demands. Each time an occupation is made, the corresponding termination rate is set equal to 2σ or $2(1-\sigma)$ with probabilities σ and $1-\sigma$, respectively.

Now, let the arrival rate of the Poisson input process be ϱ, let there be one server and an infinite queue. It is required to determine the average waiting-time for delayed demands.

We introduce the following complete set of states of the system and their probabilities (stationarity assumed):

State	Description	Probability
[–]	system empty	p
$[A, r]$	type A occupation, r in queue	f_r
$[B, r]$	type B occupation, r in queue	g_r

The transition scheme is (σ' denotes $1-\sigma$):

[–] ⇄ [A,0]: $\sigma\rho$ (to [A,0]), 2σ (to [–])
[–] ⇄ [B,0]: $\sigma'\rho$ (to [B,0]), $2\sigma'$ (to [–])
[A,0] ⇄ [A,1] ⇄ [A,2] ···: ρ (right), $2\sigma^2$ (left)
[B,0] ⇄ [B,1] ⇄ [B,2] ···: ρ (right), $2\sigma'^2$ (left)
[A,1] → [B,0], [B,1] → [A,0], [A,2] → [B,1], [B,2] → [A,1]: $2\sigma\sigma'$

It leads to the following set of equations of state:

$$0 = -\varrho p + 2\sigma f_0 + 2(1-\sigma)\,g_0, \tag{4.5.3}$$

$$0 = -(\varrho + 2\sigma) f_r + \varrho f_{r-1} + \delta_r^0 \sigma \varrho p + 2\sigma^2 f_{r+1} + 2\sigma(1-\sigma)\,g_{r+1}, \tag{4.5.4}$$

$$0 = -\{\varrho + 2(1-\sigma)\}\,g_r + \varrho g_{r-1} + \delta_r^0 (1-\sigma)\,\varrho p + 2(1-\sigma)^2\,g_{r+1} + 2\sigma(1-\sigma) f_{r+1} \quad (r = 0, 1, \ldots). \tag{4.5.5}$$

Introduce the generating functions $f_r \triangleq F(x)$ and $g_r \triangleq G(x)$. The last three equations yield

$$\varrho p = 2\sigma F(0) + 2(1-\sigma)\,G(0), \tag{4.5.6}$$

$$0 = -(\varrho - \varrho x + 2\sigma)\,F(x) + \sigma\varrho p + 2\sigma^2\,\frac{F(x) - F(0)}{x} + 2\sigma(1-\sigma)\,\frac{G(x) - G(0)}{x}, \tag{4.5.7}$$

$$0 = -(\varrho - \varrho x + 2 - 2\sigma)\,G(x) + (1-\sigma)\,\varrho p + 2(1-\sigma)^2 \times \frac{G(x) - G(0)}{x} + 2\sigma(1-\sigma)\,\frac{F(x) - F(0)}{x}. \tag{4.5.8}$$

Insertion of (4.5.6) into (4.5.7) and (4.5.8) yields

$$\{2\sigma^2 - x(\varrho - \varrho x + 2\sigma)\}\,F(x) + 2\sigma(1-\sigma)\,G(x) = \sigma\varrho(1-x)\,p, \tag{4.5.9}$$

$$2\sigma(1-\sigma)\,F(x) + \{2(1-\sigma)^2 - x(\varrho - \varrho x + 2 - 2\sigma)\}\,G(x) = (1-\sigma)\,\varrho(1-x)\,p. \tag{4.5.10}$$

Hence, $F(x)$ and $G(x)$ may be expressed in terms of p, which in turn follows from the norming equation:

$$p + F(1) + G(1) = 1. \tag{4.5.11}$$

By differentiating the relations (4.5.9) and (4.5.10) twice with respect to x and then putting $x = 1$ in the original equations and the differentiated ones, those six relations, together with (4.5.11) yield $F(1)$, $G(1)$,

$F'(1)$, $G'(1)$, $F''(1) - G''(1)$ and p. From $F'(1)$ and $G'(1)$ we eventually find the average waiting-time for delayed demands:

$$W = \frac{1}{\varrho^2} E(\underline{w}_{\text{tot}}) = \frac{1}{\varrho^2} \sum_{r=1}^{\infty} r(f_r + g_r) = \frac{1}{\varrho^2} [F'(1) + G'(1)]$$
$$= \frac{1}{4\sigma(1-\sigma)(1-\varrho)}. \tag{4.5.12}$$

For $\sigma \neq \frac{1}{2}$ it is clear that $W > 1/(1 - \varrho)$, i.e. larger than in the exponential (1) case. It should be observed that for $\sigma \approx 0$ (or ≈ 1) the increase of the waiting-time may be considerable. Serious attention should be given to this point, as the prolonged tail in the distribution of holding-times may be hardly observable!

The hyperexponential distributions thus form a family that is to some extent complementary to the Erlang-k family. They are useful when

(i) the p.d.f. of holding-times decreases monotonically;
(ii) its variance is *larger* than for the exponential (1) case.

The results for the hyperexponential case also go back to Morse (1958).

5. NON-STATIONARY BEHAVIOUR

IN earlier chapters we have always assumed stationarity to be possible. Steady-state probabilities, etc., then have been derived under this assumption. We did so in the engineer's belief that nature would be so kind as to warn us when this procedure was wrong. This indeed was the case in Section 2.3 ($M/M/c$-delay), where stationarity was clearly impossible for $\varrho \geqq c$. In the present chapter we shall pay some attention to the justification of those procedures. Moreover, even when we take the possibility of stationarity for granted, it is a real problem to investigate how long transients persist, i.e. how long it takes before the initial situation has worn out, speaking stochastically.

In Section 5.1 we shall consider the general case of a system with a finite number of possible states, a Markov process. In Section 5.2 we shall take the case $M/M/c$-blocking as a special one. When the number of possible states is infinite, the general theory is very complicated. As an example we shall deal with the case $M/G/1$-delay in Section 5.3.

5.1. Markov process with a finite number of states

Let $[i]$ $(i = 1, \ldots, n)$ be the possible states of a system. Denote by t_{ij} $(i \neq j)$ the rate at which existing states $[i]$ change into states $[j]$ (independent of time). Finally, let π_i $(i = 1, \ldots, n)$ be the probability of the system being in $[i]$ at $\tau = 0$ $(\sum \pi_i = 1)$. It is required to determine the probability $p_i(\tau)$ $(i = 1, \ldots, n)$ for the system to be in $[i]$ at time τ.

The equations of state for the non-stationary case may be obtained as follows:

[State] and probability at time τ		Transition and probability	[State] and probability at time $\tau + \Delta\tau$
$[i]$ $p_i(\tau)$	$(i \neq j)$	$[i] \to [j]$, $t_{ij}\Delta\tau$	$[j]$ $p_j(\tau + \Delta\tau)$
$[j]$ $p_j(\tau)$		no change, $\prod_{k=1,\neq j}^{n} (1 - t_{jk}\Delta\tau)$	

Neglecting higher-order effects

$$p_j(\tau + \Delta\tau) = \sum_{i=1,\neq j}^{n} p_i(\tau) \cdot t_{ij}\Delta\tau + p_j(\tau) \cdot \left\{1 - \Delta\tau \cdot \sum_{k=1,\neq j}^{n} t_{jk}\right\}.$$

When passing to the limit $\Delta\tau \to 0$

$$\frac{dp_j}{d\tau} = \sum_{i=1,\neq j}^{n} p_i(\tau)\, t_{ij} - p_j(\tau) \cdot \sum_{k=1,\neq j}^{n} t_{jk} \quad (j = 1, \ldots, n). \qquad (5.1.1)$$

When we write

$$t_{jj} := - \sum_{k=1,\neq j}^{n} t_{jk}, \qquad (5.1.2)$$

the equation may be written as

$$\frac{dp_j}{d\tau} = \sum_{i=1}^{n} p_i(\tau)\, t_{ij} \quad (j = 1, \ldots, n). \qquad (5.1.3)$$

Together with the initial conditions

$$p_i(0) = \pi_i \quad \left(\sum_{i=1}^{n} \pi_i = 1\right) \quad (i = 1, \ldots, n) \qquad (5.1.4)$$

they define the stochastic behaviour of the system.

When we introduce matrix-vector notation,

$$\boldsymbol{p}(\tau) = (p_i(\tau)); \quad \boldsymbol{\pi} = (\pi_i); \quad T = \{t_{ij}\} \qquad (5.1.5)$$

the differential equations and initial conditions may be written as

$$\frac{d}{d\tau}\mathbf{p} = p(\tau)\,T \tag{5.1.6}$$

and

$$\mathbf{p}(0) = \boldsymbol{\pi}. \tag{5.1.7}$$

It follows from (5.1.2) that T has zero row-sums. Hence,

$$\frac{d}{d\tau}\sum_1^n p_i = 0.$$

This means that the sum of the elements of $\mathbf{p}(\tau)$ remains equal to unity.

The formal solution of (5.1.6) and (5.1.7) involves the eigenvalues $\lambda_1, \ldots, \lambda_n$ and the corresponding row-eigenvectors $\mathbf{e}_1, \ldots, \mathbf{e}_n$ of the matrix T. When $\boldsymbol{\pi}$ is decomposed into those eigenvectors

$$\boldsymbol{\pi} = \sum_1^n c_k \mathbf{e}_k \tag{5.1.8}$$

the solution is

$$\mathbf{p}(\tau) = \sum_1^n c_k \mathbf{e}_k e^{\lambda_k \tau}. \tag{5.1.9}$$

When there are eigenvalues that are equal, confluent forms of (5.1.9) occur, that contain products of exponentials and polynomials.

The normal procedure to deal with those problems is to assume stationarity and to solve the system

$$\begin{aligned} \mathbf{p}T &= \mathbf{0}, \\ \sum_1^n p_i &= 1 \end{aligned} \tag{5.1.10}$$

(cf. Section 2.2, etc.). Uniqueness of this solution is tacitly taken as a warrant for stationarity to obtain for $\tau \to \infty$. Now, stationarity obtains for $\tau \to \infty$ when (i) all non-zero eigenvalues of T have negative real parts, and (ii) there is one and not more than one zero eigenvalue (say $\lambda_1 = 0$). For in this case transients decay exponentially and we have [cf. (5.1.9)] $\mathbf{p}(\infty) = c_1\mathbf{e}_1$. As $c_1\mathbf{e}_1$ satisfies $c_1\mathbf{e}_1 = \mathbf{0}$ and the element-sum of $\mathbf{p}(\infty)$ is 1, $\mathbf{p}(\infty)$ is the unique solution of (5.1.10).

In this section we shall now give a proof of the facts that non-zero eigenvalues of T have negative real parts and that uniqueness of the solution of (5.1.10) entails unicity of the eigenvalue zero. The proof is nearly identical to Giltay's proof (1950), who used a lemma by Hadamard (1903) which is related to Gerschgorin's theorem (1931), used below.

The fact that all row-sums of T are zero proves the existence of at least one zero eigenvalue. According to a theorem of Gerschgorin every eigenvalue of an $n \times n$ matrix $A = \{a_{ij}\}$ lies (in the complex λ-plane) in at least one of the n circular discs (boundaries included) with centres a_{ii} and radii $\sum_{j, \neq i} |a_{ij}|$. For the matrix T those centres lie on the negative axis and the circles pass through $\lambda = 0$. This clearly proves that any non-zero eigenvalue must have negative real part.

A multiple eigenvalue $\lambda = 0$ is possible only in case det $(T - \lambda I)$ has a multiple factor λ, or

$$\frac{d}{d\lambda} \det (T - \lambda I) = 0 \quad \text{for} \quad \lambda = 0. \tag{5.1.11}$$

If T_{kk} is the matrix obtained by omitting the kth row and column from T, and Δ_{kk} is its determinant, (5.1.11) is equivalent to

$$\Delta_{11} + \Delta_{22} + \cdots + \Delta_{nn} = 0. \tag{5.1.12}$$

Let $T_{kk}(\beta)$ be the matrix obtained from T_{kk} by multiplying all non-diagonal elements by β. For $\beta \in [0, 1]$ it is clear that in every row of $T_{kk}(\beta)$ the sum of the non-diagonal elements (which are non-negative) is less than the absolute value of the diagonal element. Hence, all Gerschgorin circles lie to the left of the imaginary axis, and $T_{kk}(\beta)$ cannot have a zero eigenvalue. But then

$$\det T_{kk}(\beta) \neq 0 \quad \text{for} \quad \beta \in [0, 1). \tag{5.1.13}$$

As this determinant is a continuous function of β, it is clear that $\Delta_{kk} = \det T_{kk}(1)$ is either zero or has the same sign as det $T_{kk}(0)$. Now, the latter determinant equals the product of $n - 1$ negative diagonal elements; its sign is $(-)^{n-1}$. Hence, all non-zero Δ_{kk} have same sign. Consequently (5.1.12) cannot be fulfilled when one of its constituents differs from zero. Now, let (5.1.10) have a unique solution **p**. Suppose in this solution that $p_j \neq 0$. Then $\Delta_{jj} \neq 0$. Hence, (5.1.12) is not valid and there is only one zero eigenvalue of T. This proves the existence of the limiting distribution $\mathbf{p}(\infty) = \mathbf{p}$, independent of the initial condition. The Markov process is then called *completely ergodic*. This is the usual case. When the process is not completely ergodic there are at least two independent eigenvectors with zero eigenvalue. The corresponding coefficients in (5.1.8) will generally depend on $\boldsymbol{\pi}$, showing that now the limiting distribution $\mathbf{p}(\infty)$ will depend on the initial condition. It may easily be shown that the system contains at least two so-called *recurrent chains*. A recurrent chain is a subset of the set of states, that does not possess possible transitions from one of its states to states outside the subset (cf. Fig. 5.1), as for instance the subsets $\{1,2,3\}$ and $\{5\}$ in Fig. 5.1.

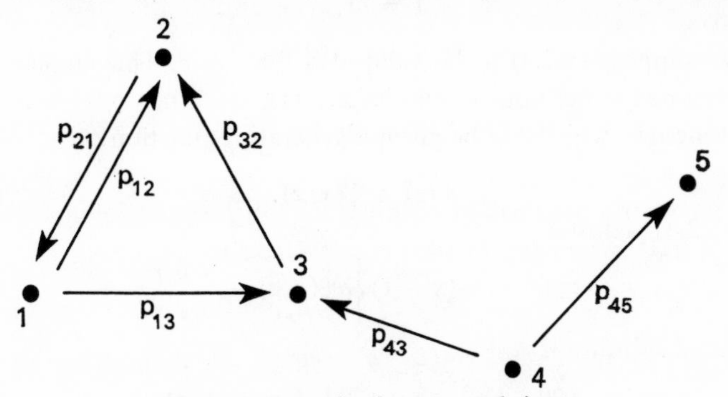

FIG. 5.1. Example of recurrent chains.

5.2. The case *M*/*M*/*c*-blocking; transient behaviour

Let $p_r(\tau)$ be the probability of r occupied servers at time τ. The rates of transition are those given in Section 2.1. Then the set of eqns. (5.1.1) in this case is

$$\frac{dp_r}{d\tau} = \varrho p_{r-1}(\tau) - (\varrho + r)\, p_r(\tau) + (r+1)\, p_{r+1}(\tau) \quad (p_{-1} \equiv 0;\ r = 0, \ldots, c-1). \tag{5.2.1}$$

$$\frac{dp_c}{d\tau} = \varrho p_{c-1}(\tau) - c p_c(\tau). \tag{5.2.2}$$

We shall deal with a special case. It will be assumed that at $\tau = 0$ $p_0(0) = \cdots = p_{c-1} = 0$; $p_c(0) = 1$. So at the beginning all c servers are engaged. The quantity $p_c(\tau)$—which we shall call the *conditional probability of blocking*—then indicates how rapidly the influence of this information will fade away. Ultimately $p_c(\tau)$ is expected to return to the unconditional blocking probability $B = p_c$ given by (2.1.9).

Introduce the Laplace Transforms:

$$p_r(\tau) \doteqdot \mathfrak{p}_r(z) \quad (r = 0, \ldots, c). \tag{5.2.3}$$

Then (5.2.1) and (5.2.2) together with initial conditions yield

$$z\mathfrak{p}_r = \varrho\mathfrak{p}_{r-1} - (\varrho + r)\,\mathfrak{p}_r + (r+1)\,\mathfrak{p}_{r+1} \quad (r = 0, \ldots, c-1), \tag{5.2.4}$$

$$z\mathfrak{p}_c - 1 = \varrho\mathfrak{p}_{c-1} - c\mathfrak{p}_c. \tag{5.2.5}$$

Now, suppose (5.2.4) to be valid also for $r \geqq c$. The supplementary equations define fictitious quantities $\mathfrak{p}_{c+1}(z)$, etc. Then $\mathfrak{p}_r(z)$ is an arithmetic function, which can be given a generating function:

$$\mathfrak{p}_r(z) \triangleq \mathfrak{P}(x, z). \tag{5.2.6}$$

Then (5.2.4) yields

$$z\mathfrak{P}(x, z) = (x - 1)\left\{\varrho\mathfrak{P}(x, z) - \frac{\partial\mathfrak{P}}{\partial x}\right\}. \tag{5.2.7}$$

Its general solution is

$$\mathfrak{P}(x, z) = C(z)\,\mathrm{e}^{-\varrho(1-x)}(1 - x)^{-z}. \tag{5.2.8}$$

The corresponding arithmetic function may be expressed in a "generalized zth sum" (cf. A 20):

$$\mathfrak{p}_r(z) = C(z)\,\varphi_r^z. \tag{5.2.9}$$

Now, $\mathfrak{p}_{c-1}$ and $\mathfrak{p}_c$ still have to satisfy (5.2.5). This yields

$$C(z) = 1/[z\varphi_c^z - \varrho\varphi_{c-1}^z + c\varphi_c^z].$$

The denominator may be simplified by the use of (A 17′ and 18):

$$[\ldots] = z\varphi_c^z + (c\varphi_c^z - \varrho\varphi_{c-1}^z) = z(\varphi_c^z + \varphi_{c-1}^{z+1}) = z\varphi_c^{z+1}.$$

So $C(z)$ has been found and

$$\mathfrak{p}_r(z) = \varphi_r^z / z\varphi_c^{z+1}. \tag{5.2.10}$$

Especially

$$p_c(\tau) \fallingdotseq \mathfrak{p}_c(z) = \varphi_c^z / z\varphi_c^{z+1}. \tag{5.2.11}$$

This then is the Laplace Transform of the conditional probability of blocking. With the aid of (A 20) we can write

$$p_c(\tau) \fallingdotseq \frac{\left[\frac{\varrho^c}{c!} + \frac{z}{1!} \frac{\varrho^{c-1}}{(c-1)!} + \frac{z(z+1)}{2!} \frac{\varrho^{c-2}}{(c-2)!} + \cdots + \frac{z(z+1)\cdots(z+c-1)}{c!}\right]}{z\left[\frac{\varrho^c}{c!} + \frac{z+1}{1!} \frac{\varrho^{c-1}}{(c-1)!} + \frac{(z+1)(z+2)}{2!} \frac{\varrho^{c-2}}{(c-2)!} + \cdots + \frac{(z+1)(z+2)\cdots(z+c)}{c!}\right]} \tag{5.2.12}$$

The denominator is of degree $c + 1$ in z, the numerator of degree c. As can be shown by a process of general induction (Haantjes, 1938), the denominator possesses $c + 1$ zeros $z_0, \ldots, z_c$ (one of them is 0, say $z_0 = 0$), which are:

(i) all different;
(ii) real and non-positive;
(iii) separated by intervals that are larger than unity; this property is not used itself, but is necessary for carrying out the induction.

Hence, (5.2.12) may be split up into simple partial fractions with denominators z and $z - z_i$ ($i = 1, \ldots, c$; $z_i < 0$):

$$p_c(\tau) \fallingdotseq \frac{A_0}{z} + \sum_{i=1}^{c} \frac{A_i}{z - z_i}. \tag{5.2.13}$$

The original function then is

$$p_c(\tau) = A_0 + \sum_{i=1}^{c} A_i e^{z_i \tau}. \tag{5.2.14}$$

The value for $\tau = 0$ may be obtained as follows (cf. Appendix and 5.2.12):

$$p_c(0) = \lim_{z \to \infty} z\mathfrak{p}_c(z) = 1. \tag{5.2.15}$$

From (5.2.14) it is then clear that $p_c(\tau)$ decreases according to the attenuating exponential terms of the sum. Finally it reaches the value A_0. According to the theory of decomposition into partial fractions we have (cf. 5.2.11)

$$A_0 = \lim_{z \to 0} z\mathfrak{p}_c(z) = \varphi_c/\varphi_c^1. \tag{5.2.16}$$

This is the correct limiting value (cf. 2.1.9).

The speed at which this value is obtained is, roughly speaking, determined by the value of z_1 (i.e. the coefficient of τ in the exponent of the

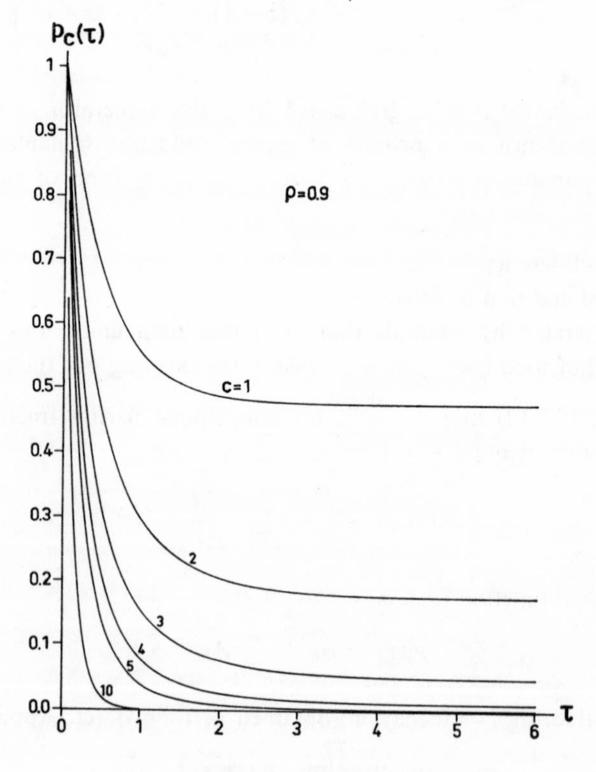

FIG. 5.2. Conditional probability of blocking.

slowest extinguishing term). Now, z_1 is the least negative zero of φ_c^{1+z}. Using the values φ_c^1 and φ_c for $z = 0$ and $z = -1$, respectively, we find a first approximation of z_1 by linear extrapolation:

$$z_1 \approx -\varphi_c^1/(\varphi_c^1 - \varphi_c) = -\varphi_c^1/\varphi_{c-1}^1.$$

This is less than -1. Hence the attenuating terms in (5.2.14) decay more rapidly than according to $e^{-\tau}$.

In Fig. 5.2 $p_c(\tau)$ has been drawn for a number of cases. The curves have *not* been obtained by the foregoing analysis, but by straightforward integration of eqns. (5.2.1) and (5.2.2) by computer.

5.3. Transient behaviour of the system $M/G/1$-delay

It is the aim of this section to investigate the transient behaviour of the system $M/G/1$ that starts "empty" at $\tau = 0$.

Introduce the following states and their time-dependent probabilities:

State	Description	Probability at time u
[−]	system empty	$p(u)$
$[r, \tau, d\tau]$	server occupied; occupation began $\tau(+d\tau)$ ago; r in queue $(r = 0, 1, 2, \ldots)$	$q_r(\tau, u)\, d\tau$

The equations of state are obtained as follows. We investigate how the states $[-]$, $[r, \tau + \Delta u, d\tau]$ and $[r, 0, \Delta u]$ at time $u + \Delta u$ can have been formed during the last interval Δu. These cases will be called A, B and C, respectively. In case B the occupation at $u + \Delta u$ was already present at time u, whilst in C this has started during $(u, u + \Delta u)$. The scheme of transitions is as follows:

	[State] and probability at time u	Transition and probability	[State] and probability at time $u + \Delta u$
(A)	$[-]$ $p(u)$	no demand $\longrightarrow$ $1 - \varrho\Delta u$	$[-]$ $p(u + \Delta u)$
	$[0, \tau, d\tau]$ $q_0(\tau, u)\, d\tau$ (all τ)	end of occupation $\longrightarrow$ $g(\tau)\Delta u$	
(B)	$[r, \tau, d\tau]$ $q_r(\tau, u)\, d\tau$	nothing $\longrightarrow$ $\{1 - g(\tau)\Delta u\}(1 - \varrho\Delta u)$	$[r, \tau + \Delta u, d\tau]$ $q_r(\tau + \Delta u, u + \Delta u)\, d\tau$
	$[r - 1, \tau, d\tau]$ $q_{r-1}(\tau, u)\, d\tau$	new demand $\longrightarrow$ $\varrho\Delta u$	
(C)	$(r + 1, \tau, d\tau]$ $q_{r+1}(\tau, u)\, d\tau$ (all τ)	end of occupation $\longrightarrow$ $g(\tau)\Delta u$	$[r, 0, \Delta u]$ $q_r(0, u + \Delta u)\, \Delta u$
	$[-]$ $p(u)$	new demand ($r = 0$ only) $\longrightarrow$ $\varrho\Delta u \cdot \delta_r^0$	

For (A) we obtain in the usual way:

$$\frac{dp}{du} = -\varrho p + \int_0^\infty q_0(\tau, u)\, g(\tau)\, d\tau. \tag{5.3.1}$$

For (B):

$$q_r(\tau + \Delta u, u + \Delta u) = q_r(\tau, u)\, d\tau \cdot \{1 - g(\tau)\,\Delta u\}\,(1 - \varrho\Delta u) + q_{r-1}(\tau, u)\, d\tau \cdot \varrho\Delta u$$

or when passing to the limit $\Delta u \to 0$:

$$\frac{\partial q_r}{\partial \tau} + \frac{\partial q_r}{\partial u} = -\{\varrho + g(\tau)\}\, q_r + \varrho q_{r-1}. \tag{5.3.2}$$

And for (C):

$$q_r(0, u) = \int_0^\infty q_{r+1}(\tau, u)\, g(\tau)\, d\tau + \varrho p(u)\, \delta_r^0. \quad (5.3.3)$$

Let us assume that at $u = 0$ the system starts empty:

$$p(0) = 1; \quad q_r(\tau, 0) = 0. \quad (5.3.4)$$

Let $\mathfrak{p}(z)$ and $\mathfrak{q}_r(\tau, z)$ be the Laplace Transforms with respect to $u \to z$,

$$p(u) \risingdotseq \mathfrak{p}(z); \quad q_r(\tau, u) \risingdotseq \mathfrak{q}_r(\tau, z). \quad (5.3.5)$$

Then (5.3.1), (5.3.2) and (5.3.3) together with the initial conditions (5.3.4) yield

$$z\mathfrak{p}(z) - 1 = -\varrho\mathfrak{p}(z) + \int_0^\infty \mathfrak{q}_0(\tau, z)\, g(\tau)\, d\tau, \quad (5.3.6)$$

$$\frac{\partial \mathfrak{q}_r}{\partial \tau} + z\mathfrak{q}_r = -\{\varrho + g(\tau)\}\, \mathfrak{q}_r + \varrho \mathfrak{q}_{r-1}, \quad (5.3.7)$$

$$\mathfrak{q}_r(0, z) = \int_0^\infty \mathfrak{q}_{r+1}(\tau, z)\, g(\tau)\, d\tau + \varrho\mathfrak{p}(z)\, \delta_r^0. \quad (5.3.8)$$

Now, the generating function $\mathfrak{Q}(x, \tau, z)$ of $\mathfrak{q}_r(\tau, z)$ is introduced:

$$z\mathfrak{p}(z) - 1 = -\varrho\mathfrak{p}(z) + \int_0^\infty \mathfrak{Q}(0, \tau, z)\, g(\tau)\, d\tau, \quad (5.3.9)$$

$$\frac{\partial \mathfrak{Q}}{\partial \tau} + z\mathfrak{Q} = \{\varrho x - \varrho - g(\tau)\}\, \mathfrak{Q}, \quad (5.3.10)$$

$$\mathfrak{Q}(x, 0, z) = \frac{1}{x} \int_0^\infty \{\mathfrak{Q}(x, \tau, z) - \mathfrak{Q}(0, \tau, z)\}\, g(\tau)\, d\tau + \varrho\mathfrak{p}(z). \quad (5.3.11)$$

The solution of (5.3.10) is

$$\mathfrak{Q}(x, \tau, z) = K(x, z) \exp\{-(\varrho - \varrho x + z)\tau - G(\tau)\}. \quad (5.3.12)$$

Substitution of (5.3.9) and (5.3.12) into (5.3.11) yields a relation between $K(x, z)$ and $K(0, z)$, which may be written as

$$K(x, z) = \frac{(\varrho x - \varrho - z)\, K(0, z)\, J(0, z) + \varrho x}{(\varrho + z)\, \{x - J(x, z)\}}, \quad (5.3.13)$$

where J is defined by

$$J(x, z) := \int_0^\infty \exp\{-\varrho(1 - x)\tau - z\tau\} f(\tau)\, d\tau. \qquad (5.3.14)$$

The probability that the server is occupied at time u, whilst the number of waiting demands is r, equals $\int_0^\infty q_r(\tau, u)\, d\tau$. Hence, this integral must be less than 1. In connection with the fact that $g(\tau)$ is bounded for distributions from the CM-class (say less than m), it then follows from (5.3.3) that $q_r(0, u)$ is less than $M := m + \varrho$. Consequently, $\mathfrak{q}_r(0, z) < M/\delta$ when $\operatorname{Re} z > \delta > 0$. Hence,

$$\sum_{r=0}^{\infty} \mathfrak{q}_r(0, z)\, x^r$$

must converge for $x < 1$ and $\operatorname{Re} z > \delta > 0$. This means that $K(x, z) = \mathfrak{Q}(x, 0, z)$ cannot have a pole for $\operatorname{Re} z > 0$ when $|x| < 1$. Now, the expression between $\{\ \}$ in the denominator of (5.3.13) has a zero for $x = \xi(z)$, defined implicitly by

$$\xi(z) = \int_0^\infty \exp[-\varrho\{1 - \xi(z)\}\tau - z\tau] f(\tau)\, d\tau. \qquad (5.3.15)$$

The pair $z = 0$, $\xi(z) = 1$ satisfies (5.3.15). Now, suppose $\xi(z) = 1 + \alpha z + O(z^2)$. Then (5.3.15) yields

$$1 + \alpha z + O(z^2) = 1 + \alpha\varrho z - z + O(z^2)$$

and hence

$$\alpha = \frac{-1}{(1 - \varrho)}; \qquad \xi(z) \approx 1 - \frac{z}{1 - \varrho} + \cdots. \qquad (5.3.16)$$

So for positive z there is a solution $\xi(z)$ of (5.3.15) that is less than 1 (for $\varrho < 1$). In order that $K(x, z)$ remain finite when $z > 0$ and $x = \xi(z)$, the numerator of the expression (5.3.13) should vanish at the same time:

$$\{\varrho\xi(z) - \varrho - z\}\, K(0, z)\, J(0, z) + \varrho\xi(z) = 0. \qquad (5.3.17)$$

From (5.3.17), (5.3.13), (5.3.12) and (5.3.14) we obtain

$$\int_0^\infty \mathfrak{Q}(0, \tau, z)\, g(\tau)\, d\tau = \frac{\varrho\xi(z)}{\varrho + z - \varrho\xi(z)}.$$

Insertion into (5.3.9) finally yields

$$\mathfrak{p}(z) = \frac{1}{\varrho + z - \varrho\xi(z)}, \tag{5.3.18}$$

which is the Laplace Transform of the time-dependent probability $p(u)$ of finding the system empty. Its complement $1 - p(u)$ is the time-dependent probability of delay.

Apart from simple cases like $M/M/1$ it is not possible to find the original function $p(u)$ in an analytical way. We shall concentrate here on the asymptotic behaviour for $u \to \infty$.

For $\mathfrak{p}(z)$ to have a pole, the denominator in (5.3.18) should vanish. From (5.3.15) it ensues that in that case $\xi(z) = 1$. Then (5.3.18) shows that the denominator equals z. Hence, $z = 0$ is the only pole of $\mathfrak{p}(z)$.

Now we observe that

$$\lim_{z\to 0} z\mathfrak{p}(z) = \lim_{z\to 0} \frac{z}{\varrho + z - \varrho\left(1 - \dfrac{z}{1-\varrho} + \cdots\right)} = 1 - \varrho.$$

If we assume $\lim_{u\to\infty} p(u)$ to exist, it must have this value $1 - \varrho$, which we know to be correct.

The asymptotic way in which $p(u)$ attains the value $1 - \varrho$ depends on the other singularities of $\mathfrak{p}(z)$ (in the left half-plane). These can be branch-points. For $\xi(z)$ (and hence $\mathfrak{p}(z)$) to have a branch-point in z, $dz/d\xi$ should be zero. Differentiate (5.3.15):

$$1 = \left(\varrho - \frac{dz}{d\xi}\right) \int_0^\infty \exp\left[-\varrho\{1 - \xi(z)\}\,\tau - z\tau\right] \tau f(\tau)\, d\tau.$$

For a branch-point z:

$$\frac{1}{\varrho} = \int_0^\infty \exp\left[-\varrho\{1 - \xi(z)\,\tau\} - z\tau\right] \tau f(\tau)\, d\tau. \tag{5.3.19}$$

As an example we shall take the case of an Erlang-k distribution:

$$f(\tau) = k^k \frac{\tau^{k-1}}{(k-1)!} e^{-k\tau}.$$

Then the right-hand sides of (5.3.15) and (5.3.19) can be evaluated:

$$\xi(z) = \left[\frac{k}{\varrho\{1 - \xi(z)\} + z + k}\right]^k, \tag{5.3.20}$$

$$\frac{1}{\varrho} = \left[\frac{k}{\varrho\{1 - \xi(z)\} + z + k}\right]^{k+1}. \tag{5.3.21}$$

By elimination of $\xi(z)$

$$z = -\varrho - k + (1 + k)\,\varrho^{1/(k+1)} e^{2m\pi i/(k+1)} \quad (m = 0, \ldots, k). \tag{5.3.22}$$

These are $k + 1$ branch-points, equidistant on a circle with centre $-\varrho - k$ and radius $(1 + k)\,\varrho^{1/(1+k)}$.

The asymptotic behaviour is mainly determined by the rightmost of those branch-points, i.e.

$$z_0 = -\varrho - k + (1 + k)\,\varrho^{1/(k+1)}. \tag{5.3.23}$$

Apart from a fractional power of u, this transient is characterized by an exponential factor $e^{z_0 u}$. The value of z_0 then yields an indication about the rate of convergence to the limit $u \to 1 - \varrho$. For $k = 1, 2, 3, \infty$ and $\varrho = 0.5, 0.6, 0.7, 0.8, 0.9$ one obtains for $-1/z_0$:

$-1/z_0 =$

ϱ \ k	1	2	3	∞
0.5	11.7	8.4	7.3	5.2
0.6	19.7	14.3	12.6	9.0
0.7	37.5	27.6	24.3	17.6
0.8	89.7	66.5	58.7	43.2
0.9	380	283	251	187

For large values of ϱ the values of $-1/z_0$, which roughly determine the decay-times of the transient, may be quite considerable. They do, however, not depend heavily on the p.d.f. $f(\tau)$. It is an open question whether this is also true when other types of distributions (e.g. hyperexponential or bimodal cases) are considered.

When $k = 1$, $\xi(z)$ may be obtained algebraically from (5.3.20)

$$\xi(z) = \frac{\varrho + z + 1}{2\varrho} - \sqrt{\left(\frac{\varrho + z + 1}{2\varrho}\right)^2 - \frac{1}{\varrho}}, \qquad (5.3.24)$$

and hence, from (5.3.18),

$$\frac{1}{z} - \mathfrak{p}(z) = \frac{\varrho}{z}\left\{\frac{\varrho + z + 1}{2\varrho} - \sqrt{\left(\frac{\varrho + z + 1}{2\varrho}\right)^2 - \frac{1}{\varrho}}\right\}. \qquad (5.3.25)$$

Now, the original function of the left-hand member is the complement of the probability of an empty system at time τ, i.e. the probability $D(u)$ of delay for a demand arriving at time u. As the right-hand side equals ϱ/z times the expression in (3.2.7) for $c = 1$, it follows that for the $M/M/1$-delay system $D(u)$ equals ϱ times the c.d.f. of waiting-times under "last-come-first-served" conditions.

6. PRIORITY

CONSIDER a $M/G/1$-delay system with n groups of sources: $i = 1, ..., n$. Each group produces its own Poisson arrival process, with its own arrival rate. Demands from source group i—said to be *class i demands*—lead to *class i occupations*. Each occupation class has its own holding-time distribution.

Waiting demands of class i will receive service prior to those of class $i+1, ..., n$. It is—alas!—standard practice to use a notation which attributes "higher" priority to a class with lower index i. In case of arrival of higher priority demands at a moment at which a lower priority item is being served, there are two possibilities. In the case of *non-pre-emptive priority* any occupation is first finished regardless of the priority. When the occupation comes to an end the server turns to helping (one of) the waiting demand(s) with highest priority. In the case of *pre-emptive priority*, however, any occupation is interrupted as soon as a demand of higher priority arrives. The higher-order demand is given service and the interrupted item goes back to (the head of) the queue. It remains there until the server has dealt with the interrupting demand and all other higher priority demands that have arrived since the interruption until ultimately there are no higher priority demands any more. There are various forms of pre-emptive priority discipline, according to what happens when the server recommences the serving of the partially served demand, viz.

(i) *Pre-emptive-resume*: the helping is resumed where it was interrupted. No serving time is lost. Hence, the time spent by the server in further helping the interrupted demand is equal to the original scheduled holding-time minus all lapses of time spent in servicing it in the past,

(ii) *Pre-emptive-repeat without resampling*: now the helping time before the interruption is lost. After the interruption the total holding-time, as scheduled before the interruption, must again be spent by the server.

(iii) *Pre-emptive-repeat with resampling*: again the time before the interruption is lost, but the new holding-time is a *new* sample from the holding-time distribution.

It is not difficult to imagine situations that meet one of the conditions (i) or (ii). Condition (iii) is, for example, present when the variability of holding-times is not caused by different work specifications for the various demands, but by external factors such as the working speed being dependent on weather conditions. When holding-times are exponential the conditions (i) and (ii) are identical; in the case of constant holding-times (ii) and (iii) coincide.

It is impossible to describe all possibilities fully. In Section 6.1 we shall discuss the non-pre-emptive priority case (i). In Section 6.2 we shall deal with some simple cases of the pre-emptive type related to breakdown of machines.

6.1. Non-pre-emptive priority

We shall deal here with the general non-pre-emptive priority case. There are n priority classes $1, \ldots, n$ ($1 =$ highest priority). Each class has its own holding-time distribution, possibly with different averages. Hence, norming the time is of no value and will not be used.

It is required to compute the average waiting-times for the various classes as well as the overall average. Stationarity is assumed to exist. In order for this to be possible, ϱ should be less than 1.

A very straightforward combinatorial theory has been given by Cobham (1954, 1955). Though it is possible to use birth-and-death equations, we shall follow a combinatorial method in the present case.

Let the following be defined for class i traffic:

λ_i the arrival rate,

$\underline{h}_i$ the holding-time,

$h_i := E(\underline{h}_i)$ the average holding-time,

$\varrho_i := \lambda_i h_i$ the server occupancy,

$E(\underline{h}_i^2)$ the second moment of the distribution of holding-time;

and for the total traffic:

$\lambda = \sum_1^n \lambda_i$ the arrival rate,

$\underline{h}$ the holding-time,

$h = E(\underline{h}) = \sum_1^n \frac{\lambda_i}{\lambda} h_i = \frac{1}{\lambda} \sum_1^n \varrho_i$ the average holding-time,

$\varrho = \sum_1^n \varrho_i = \lambda h$ the server occupancy,

$E(\underline{h}^2) = \sum_1^n \frac{\lambda_i}{\lambda} E(\underline{h}_i^2)$ the second moment of the distribution of holding time.

Consider a virtual demand of priority class p. Its waiting-time $\underline{w}_p$ consists of three components:

(i) the residual duration of the occupation encountered, if any;
(ii) the sum of the holding-times of the demands of priority $1, \ldots, p$ that were present in the queue at the arrival of the virtual demand;
(iii) the sum of the holding-times of the demands of priorities $1, \ldots, p - 1$ that arrive during the waiting-time $\underline{w}_p$ itself.

The average waiting-time w_p for class p demands consists of the averages of those three components, which will be considered in turn.

There is a probability ϱ for the virtual demand to meet with delay. If so, the occupation encountered may be of any class. Its expected residual duration is $E(\underline{h}^2)/2h$ (cf. 1.2.9). Hence, the average duration under (i) is

$$\varrho E(\underline{h}^2)/2h = \tfrac{1}{2}\lambda E(\underline{h}^2).$$

The average number of class i demands in the queue, met by the virtual demand on arrival, equals the average total waiting-time of class i demands per unit of time, i.e. $\lambda_i w_i$. The average durations of the occupations ensuing from those demands are independent of their number. Hence, the expected total duration of (ii) is:

$$\sum_1^p \lambda_i w_i \cdot h_i = \sum_1^p \varrho_i w_i .$$

The average number of class i demands arising during the waiting-time $\underline{w}_p$ is $\lambda_i \underline{w}_p$. The average of the ensuing holding-times is $\lambda_i \underline{w}_p \cdot h_i = \varrho_i \underline{w}_p$. When we take the classes $i = 1, ..., p - 1$ together and take the average of $\underline{w}_p$ we obtain for (iii):

$$w_p \sum_1^{p-1} \varrho_i .$$

Hence,

$$w_p = \tfrac{1}{2}\lambda E(\underline{h}^2) + \sum_1^p \varrho_i w_i + w_p \sum_1^{p-1} \varrho_i$$

or

$$w_p\left(1 - \sum_1^p \varrho_i\right) = \tfrac{1}{2}\lambda E(\underline{h}^2) + \sum_1^{p-1} \varrho_i w_i \quad (p = 1, ..., n). \tag{6.1.1}$$

When $p = 1$ this yields

$$w_1 = \frac{\lambda E(\underline{h}^2)}{2(1 - \varrho_1)} . \tag{6.1.2}$$

The relation (6.1.1) then yields successively $w_2, ..., w_n$. Writing down (6.1.1) with p replaced by $p - 1$, and subtracting this relation from (6.1.1) yields

$$w_p\left(1 - \sum_1^p \varrho_i\right) - w_{p-1}\left(1 - \sum_1^{p-1} \varrho_i\right) = \varrho_{p-1} w_{p-1},$$

or

$$w_p = \frac{1 - \sum_1^{p-2} \varrho_i}{1 - \sum_1^p \varrho_i} \cdot w_{p-1} . \tag{6.1.3}$$

Multiple application of this recursion relation together with (6.1.2) then yields

$$w_p = \frac{\lambda E(\underline{h}^2)}{2(1 - \sum_1^p \varrho_i)(1 - \sum_1^{p-1} \varrho_i)}. \tag{6.1.4}$$

This is the average waiting-time for class i demands.

For the deduction of (6.1.4) the mutual priorities of the classes 1, ..., $p - 1$ as well as those of the classes $p + 1$, ..., n are immaterial. Hence, the foregoing analysis could have been made using three classes only.

When there is one class only (no priority) (6.1.4) reduces to

$$w = \frac{\lambda E(\underline{h}^2)}{2(1 - \varrho)}. \tag{6.1.5}$$

The average waiting-time W for delayed demands is $1/\varrho$ times as large

$$W = \frac{E(\underline{h}^2)}{2(1 - \varrho)\, h} \tag{6.1.6}$$

which results are in agreement with (4.2.22) (in the present case no time-scaling has been applied).

The overall average of waiting-time for all n classes is

$$w_{\text{overall}} = \frac{E(\underline{h}^2)}{2} \sum_{i=1}^{n} \frac{\lambda_i}{(1 - \sum_1^{i-1} \varrho_j)(1 - \sum_1^{i} \varrho_j)}. \tag{6.1.7}$$

When the priorities of two adjacent classes are interchanged ($\lambda_k \rightleftarrows \lambda_{k+1}$, $\varrho_k \rightleftarrows \varrho_{k+1}$) this only affects the terms with index $i = k$ and $i = k + 1$ in the (main) sum of (6.1.7). It is a matter of some tedious though simple algebra to prove that this interchange reduces w_{overall} if and only if $h_k > h_{k+1}$. As long as *any* $h_k > h_{k+1}$, improvement is possible. Hence, the least overall waiting-time is incurred when the classes are sequenced according to increasing average holding-time (smallest average has highest priority). The effect of this sequencing will be demonstrated by taking two classes with different constant holding-times $h_1 = 1/2\sigma$ and $h_2 = 1/2(1 - \sigma)$. Let $\lambda_1 = \sigma\lambda$ and $\lambda_2 = (1 - \sigma)\lambda$ be the arrival rates of the two classes. The average holding-time then is equal to $(\lambda_1 h_1 + \lambda_2 h_2)/\lambda = 1$.

When no priorities are assigned the average waiting-time w^* is (cf. (4.2.22); $h = 1$):

$$w^* = \frac{\lambda E(\underline{h}^2)}{2(1-\lambda)}.$$

When, however, class 1 has priority over 2, we eventually find:

$$w_1 = \frac{\lambda E(\underline{h}^2)}{2(1-\lambda/2)},$$

$$w_2 = \frac{\lambda E(\underline{h}^2)}{2(1-\lambda/2)(1-\lambda)},$$

$$w_{\text{overall}} = \frac{\lambda E(\underline{h}^2)}{2(1-\lambda)} \cdot \frac{1-\sigma\lambda}{1-\lambda/2}.$$

It is evident that the high (low) priority class experiences a smaller (larger) average waiting-time than in the case without priority. Application of priority reduces (augments) overall waiting-time whenever $h_1 < h_2$ $(h_1 > h_2)$. This is in accordance with the stated sequencing rule.

The number of priority classes need not be finite. As an example consider the case where the holding-times of arriving demands are exponential (1) and where, moreover, *those holding-times are exactly known on arrival.* Waiting demands are selected for servicing according to the least holding-time (SPT rule: "Shortest Processing Time" first). The arrival rate is ϱ. The "class" with holding-time $\tau(+d\tau)$ possesses an arrival rate $d\lambda = \varrho e^{-\tau} d\tau$. In (6.1.7) we now should take integration over τ from 0 to ∞ instead of summing from 1 to n. Moreover, instead of

$$\sum_{j=1}^{i(-1)} \varrho_j$$

we should use

$$\int_{(0)}^{(\tau)} \tau\, d\lambda = \varrho \int_0^{\tau} u e^{-u} du = \varrho(1 - e^{-\tau} - \tau e^{-\tau}).$$

Observing that $E(\underline{h}^2) = 2$ for the exponential (1) distribution, (6.1.7) transforms into

$$w_{\text{overall}} = \varrho \int_0^\infty \frac{e^{-\tau} d\tau}{\{1 - \varrho(1 - e^{-\tau} - \tau e^{-\tau})\}^2} . \tag{6.1.8}$$

This result is due to Phipps (1956). The value of this expression is increased by suppressing the term $\tau e^{-\tau}$. The resulting expression then is equal to $\varrho/(1 - \varrho)$, which is the average waiting-time without priorities. Hence, introduction of priorities is beneficial, *the more the larger* ϱ *is.* The latter statement is illustrated by Fig. 6.1.

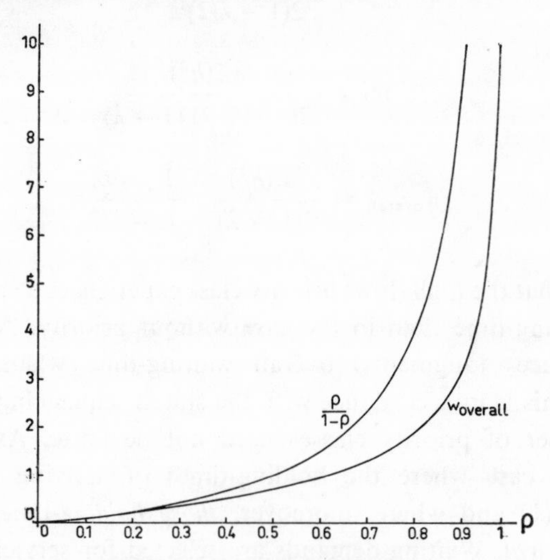

FIG. 6.1. Influence of SPT-rule.

An excellent treatment of the case at hand has been presented by Kesten and Runnenburg (1957). Cf. also Jaiswal (1968) and Conway *et al.* (1967).

6.2. Pre-emptive priority; breakdown of machines

The case of pre-emptive priority is essentially more difficult than non-pre-emptive priority, when dealt with by the method of birth-and-death equations. Apart from the case "pre-emptive-repeat with resampling", past durations of all partially finished (pre-empted) occupations must be recorded. As pre-emption may take place in succession (demands that have pre-empted lower priority occupations may in turn be pre-empted by still higher priority demands) this normally leaves us with the obligation of introducing quite a lot of supplementary variables in order to achieve "markovization".

Now, let us consider combinatorial methods. An advantage in analysis then is that the behaviour of classes $p + 1, \ldots, n$ is without any influence on the traffic of classes $1, \ldots, p$. There are also difficulties, however. One is of a conceptual nature. There are several types of interval to be considered: waiting-time (before a demand proceeds for the first time to the server), holding-time, interruption time due to pre-emption, etc. Another difficulty stems from the fact that class occupancies are not known beforehand (in any case in the pre-emptive-repeat cases) as holding-times are increased by the effect of repetition. A much more subtle analysis is needed in the case of pre-emption, resulting in rather unwieldy formulae. For a comprehensive treatment cf. Conway *et al.* (1967).

Machine breakdown and subsequent repairs may be considered a typical example of pre-emptive priority. Instead of speaking of "demands" we shall now speak about "jobs". The actual jobs form second-class priority jobs. The machine breakdowns and their subsequent repairs are first-class jobs with pre-emptive priority. The actual jobs arrive according to a Poisson process at rate λ, forming a queue before a single machine (the server). The (original) *net-processing-time* $\underline{n}$ (i.e. when uninterrupted) is supposed to be constant (1). So the system is $M/D/1$-delay. Now, the machine is assumed to be liable to breakdowns, which occur according to a Poisson process at density ν, *but only as long as the machine is working*. The repair-time following the breakdown is constant (T). We may consider two cases: (i) the time spent in helping before the occurrence of the breakdown is *not* lost (*repair-resume*) and (ii) the helping has to start anew after repair (*repair-repeat*).

We shall assume that the machine, after a breakdown and successive repair, resumes or repeats the helping of that job with which the machine was occupied before breakdown.

On arrival jobs may go through the following phases within the system. First there may be a time lag between arrival and the instant of the first entry to the machine. This will be called the *waiting-time* ($\underline{w}$). Then the job is processed by the machine. The processing may be interrupted one or more times. The total duration of the interval(s) in which the job is being processed is called the *(gross) processing-time* ($\underline{p}$). For the repair-resume case $\underline{p} = \underline{n}$; for repair-repeat $\underline{p} \geqq \underline{n}$. The total duration of the breakdown intervals (repairs) per job is called the *total breakdown-time* ($\underline{b}$) per job.

It is required to calculate the averages of $\underline{w}$, $\underline{p}$ and $\underline{b}$.

First we shall give some attention to the use of equations of state, taking the simplest case repair-repeat. We introduce the following complete set of mutually exclusive states and their steady-state probabilities:

[State]	Description	Probability
[−]	system empty	p
$[r, t, dt]$	machine works; occupation began $t(+dt)$ ago; r in queue	$q_r(t)\,dt$
$[r, t, dt]'$	machine "down", breakdown began $t(+dt)$ ago; r in queue, the interrupted item included	$q_r'(t)\,dt$

The following set of equations may readily be obtained:

$$0 = -\lambda p + q_0(1), \tag{6.2.1}$$

$$\frac{dq_r}{dt} = \lambda q_{r-1} - (\lambda + \nu)\, q_r, \tag{6.2.2}$$

$$\frac{dq_r'}{d\tau} = \lambda q_{r-1}' - \lambda q_r', \tag{6.2.3}$$

$$q_r(0) = q_{r+1}(1) + q_{r+1}'(T) + \delta_r^0 \lambda p, \tag{6.2.4}$$

$$q_r'(0) = \nu \int_0^1 q_{r-1}(u)\, du, \tag{6.2.5}$$

$$p + \sum_{r=0}^{\infty} \left[\int_0^1 q_r(u)\, du + \int_0^T q_r'(u)\, du \right] = 1. \tag{6.2.6}$$

There are no essential difficulties in solving these equations. If $Q(x, \tau)$ and $Q'(x, \tau)$ are the generating functions of $q_r(\tau)$ and $q'_r(\tau)$, respectively, the equations may be transformed and yield

$$0 = -\lambda p + Q(0, 1), \tag{6.2.7}$$

$$\frac{\partial Q}{\partial t} = (\lambda x - \lambda - \nu)\, Q, \tag{6.2.8}$$

$$\frac{\partial Q'}{\partial t} = \lambda(x - 1)\, Q', \tag{6.2.9}$$

$$xQ(x, 0) = Q(x, 1) - Q(0, 1) + Q'(x, T) - Q'(0, T) + \lambda xp, \tag{6.2.10}$$

$$Q'(x, 0) = x\nu \int_0^1 Q(x, u)\, du, \tag{6.2.11}$$

$$p + \int_0^1 Q(1, u)\, du + \int_0^T Q'(1, u)\, du = 1. \tag{6.2.12}$$

The partial differential equations (6.2.8 and 6.2.9) yield Q and Q' but for unknown multiplicative functions of x. By the use of (6.2.10 and 6.2.11) those functions may be expressed in $Q(0, 1)$ and p. The latter constants then ensue from (6.2.7 and 6.2.12). Then $Q(x, t)$ and $Q'(x, t)$ may be considered known. From those functions we may obtain the fraction of the time the machine is "down":

$$P_{\text{down}} = \int_0^T Q'(1, u)\, du. \tag{6.2.13}$$

The average breakdown-time per job is

$$E(\underline{b}) = P_{\text{down}}/\lambda. \tag{6.2.14}$$

The average gross processing-time per job is

$$E(\underline{p}) = \int_0^1 Q(1, u)\, du/\lambda. \tag{6.2.15}$$

The average total of all waiting- and breakdown-times per unit of time is

$$E(\underline{w}_{tot}) = \frac{\partial}{\partial x}\left\{\int_0^1 Q(x, u)\,du + \int_0^T Q'(x, u)\,du\right\}\Bigg|_{x=1}. \qquad (6.2.16)$$

Finally, the average waiting-time per job is

$$E(\underline{w}) = \{E(\underline{w}_{tot}) - P_{down}\}/\lambda. \qquad (6.2.17)$$

Though the analysis outlined above is very straightforward, it is so unwieldy that it will not be given in full here. Fortunately, combinatorial methods prove to be more efficient in this area. When using the methods of equations of state, the states are split up by the introduction of supplementary variables to such an extent as to render the system markovian. The combinatorial methods deal as far as possible with occupations, etc., as entities. The relations between those building-stones are on the one hand deterministic and dictated by the system, on the other hand dependent on statistical laws. Whereas the methods of equations of state have more or less a bulldozer character, the combinatorial methods normally need some subtle reasoning (cf. Section 6.1) whereby the analytical effort may, however, be reduced considerably.

We consider the machine with its breakdowns and the repair facility together as one super-machine. The holding-time ($\underline{h}$) of the supermachine consists of the gross processing-time $\underline{p}$ *and* the breakdown-time $\underline{h} = \underline{p} + \underline{b}$. Let us suppose that we are able to obtain the distribution of $\underline{h}$, and hence its moments $E(\underline{h}) = h$, $E(\underline{h}^2)$, etc. The occupancy of the supermachine then is $\varrho = \lambda h$. The waiting-time per job is obtained by the use of (4.2.22) (no scaling):

$$E(\underline{w}) = \frac{\lambda E(\underline{h}^2)}{2(1 - \varrho)}. \qquad (6.2.18)$$

The calculation of the average waiting-time is reduced to the determination of the first and second moments of the distribution of the holding-time $\underline{h}$.

Let us first consider the case repair-resume which happens to be the simpler case here. The net processing-time is constant $\underline{n} = 1$. With probability

$\varphi_s = e^{-\nu}\nu^s/s!$ it will be interrupted s times, causing a breakdown-time sT. Hence the p.d.f. $f(t)$ of the holding-time $\underline{h}$ is

$$f(t) = \sum_{s=0}^{\infty} \varphi_s \delta(t - 1 - sT), \tag{6.2.19}$$

where $\delta(t - t_0)$ denotes the (improper) Dirac function which is zero everywhere except for a unit peak at $t = t_0$.

Then we obtain for the average holding-time (cf. A 10)

$$h = E(\underline{h}) = \sum_{s=0}^{\infty} \varphi_s(1 + sT) = \Phi(1) + T\Phi'(1) = 1 + \nu T, \tag{6.2.20}$$

of which a part $E(\underline{b}) = \nu T$ is due to interruption.

For the second moment one finds (cf. A 10 and 10′)

$$E(\underline{h}^2) = \sum_{s=0}^{\infty} \varphi_s(1 + sT)^2 = \sum \varphi_s + (2T + T^2) \sum s\varphi_s + T^2 \sum s(s-1)\varphi_s$$

$$= 1 + \nu(2T + T^2) + \nu^2 T^2. \tag{6.2.21}$$

The machine occupancy is

$$\varrho = \lambda h = \lambda(1 + \nu T). \tag{6.2.22}$$

Insertion of (6.2.21 and 6.2.22) in (6.2.18) yields

$$E(\underline{w}) = \frac{\lambda\{1 + \nu(2T + T^2) + \nu^2 T^2\}}{2(1 - \lambda - \lambda\nu T)}. \tag{6.2.23}$$

This is larger than the average waiting-time per job without breakdowns, which is $\lambda/2(1 - \lambda)$. When $\lambda \to 1/(1 + \nu T)$ the denominator of (6.2.23) vanishes: for this load the queue increases indefinitely and the system ceases to be workable.

We now consider the case repair-repeat. Again we consider one gross holding-time $\underline{h}$. There is a probability $e^{-\nu}$ that the service is not interrupted and that $\underline{h} = 1$ in consequence. With probability $\nu e^{-\nu u} du$ an interruption occurs at $u(+du)$, $u < 1$. The unsuccessful processing period u and the following repair-time T are lost. After this period another lapse of time $\underline{h}^*$ is necessary that has the same p.d.f. as $\underline{h}$. In order for $\underline{h}$ to have a total length t in case of interruption, $\underline{h}^*$ should have a duration $t - u - T$. Hence, we find for the p.d.f. $f(t)$ of $\underline{h}$ (and $\underline{h}^*$)

$$f(t) = e^{-\nu}\delta(t-1) + \nu \cdot \int_0^1 e^{-\nu u} f(t-u-T)\, du. \tag{6.2.24}$$

The first moment follows by integration:

$$E(\underline{h}) = \int_0^\infty t f(t)\, dt = e^{-\nu} \cdot 1 + \nu \cdot \int_0^1 e^{-\nu u} du \int_0^\infty t f(t-u-T)\, dt$$

$$= e^{-\nu} + \nu \cdot \int_0^1 e^{-\nu u} du \int_0^\infty (v + u + T) f(v)\, dv$$

$$= e^{-\nu} + \nu \cdot \int_0^1 e^{-\nu u} \{E(\underline{h}) + u + T\}\, du$$

and hence

$$h = E(\underline{h}) = \frac{e^\nu - 1}{\nu}(1 + \nu T). \tag{6.2.25}$$

This is larger than 1, as is to be expected. The gross machine occupancy (i.e. including breakdown) is

$$\varrho = \lambda h = \lambda \frac{e^\nu - 1}{\nu}(1 + \nu T). \tag{6.2.26}$$

If $T \to 0$ in (6.2.25) we obtain the average gross processing-time per job:

$$E(\underline{p}) = (e^\nu - 1)/\nu. \tag{6.2.27}$$

The remainder of h is due to breakdown (repairs):

$$E(\underline{b}) = (e^\nu - 1)\, T. \tag{6.2.28}$$

In the same way we obtain the second moment of $\underline{h}$:

$$E(\underline{h}^2) = e^{-\nu} \cdot 1^2 + \nu \cdot \int_0^1 e^{-\nu u} du \int_0^\infty (v + u + T)^2 f(v)\, dv$$

$$= e^{-\nu} + \nu \cdot \int_0^1 e^{-\nu u} \{E(\underline{h}^2) + 2(u + T)\, E(\underline{h}) + (u + T)^2\}\, du$$

$$= 1 + E(\underline{h})\, 2(e^{\nu} - 1)\, T + 2\, \frac{e^{\nu} - 1 - \nu}{\nu}$$

$$+ T^2(e^{\nu} - 1) + 2T \cdot \frac{e^{\nu} - 1 - \nu}{\nu} + 2\, \frac{e^{\nu} - 1 - \nu - \frac{1}{2}\nu^2}{\nu^2}\,. \tag{6.2.29}$$

The average job waiting-time may be obtained by inserting (6.2.26, 6.2.25 and 6.2.29) into (6.2.18).

The combinatorial methods gain further in power by using Laplace Transforms. We shall now consider again the repair-resume case under the general assumption of *arbitrary distributions* for the *net processing-time* $\underline{n}$ and for the *repair-time* $\underline{r}$. The corresponding p.d.f.'s will be denoted by $f^N(t)$ and $f^R(t)$; the Laplace Transforms (LT's) of those functions by $\mathfrak{f}^N(z)$ and $\mathfrak{f}^R(z)$.

Let $\underline{n}$ have the value u, and let there be s breakdowns during the total lapse of time u. Then the holding-time $\underline{h}$ consists of one interval of length u (p.d.f.: $\delta(t - u)$; its LT is e^{-zu}) and s intervals having p.d.f. $f^R(t)$ (its LT is $\mathfrak{f}^R(z)$). As those $s + 1$ interval lengths are uncorrelated, the LT's multiply: LT of p.d.f. $(\underline{h} \mid \underline{n} = u, s$ breakdowns$) = e^{-zu}\{\mathfrak{f}^R(z)\}^s$. The probability of s breakdowns in time u is $e^{-\nu u}(\nu u)^s/s!$. Hence, the LT of the p.d.f. $(\underline{h} \mid \underline{n} = u)$ equals

$$e^{-zu-\nu u} \sum_{s=0}^{\infty} \{\nu u \mathfrak{f}^R(z)\}^s/s! = \exp\{-zu - \nu u + \nu \mathfrak{f}^R(z)\, u\}.$$

The LT $\mathfrak{f}(z)$ of the *unconditional* p.d.f. $f(t)$ of $\underline{h}$ is obtained by integration with weight $f^N(u)$:

$$f(t) \risingdotseq \mathfrak{f}(z) = \int_0^\infty \exp\{-zu - \nu u + \nu \mathfrak{f}^R(z)\, u\} f^N(u)\, du,$$

or, observing that $\int_0^\infty e^{-vu} f^N(u)\,du = \mathfrak{f}^N(v)$,

$$f(t) \doteqdot \mathfrak{f}(z) = \mathfrak{f}^N(z + v - v\mathfrak{f}^R(z)). \tag{6.2.30}$$

Now, by virtue of (A 25):

$$\begin{aligned} z + v - v\mathfrak{f}^R(z) &= z + v - v\{1 - zE(\underline{r}) + \tfrac{1}{2}z^2E(\underline{r}^2) - \cdots\} \\ &= z\{1 + vE(\underline{r})\} - \tfrac{1}{2}vz^2E(\underline{r}^2) + \cdots. \end{aligned}$$

Then, inserting this result in (6.2.30),

$$\begin{aligned} \mathfrak{f}(z) &= 1 - zE(\underline{h}) + \tfrac{1}{2}z^2E(\underline{h}^2) - \cdots \\ &= 1 - [z\{1 + vE(\underline{r})\} - \tfrac{1}{2}vz^2E(\underline{r}^2)]\,E(\underline{n}) + \tfrac{1}{2}[\cdots]^2\,E(\underline{n}^2) - \cdots \end{aligned}$$

or, equating coefficients of powers of z,

$$E(\underline{h}) = \{1 + vE(\underline{r})\}\,E(\underline{n}), \tag{6.2.31}$$

$$E(\underline{h}^2) = vE(\underline{r}^2)\,E(\underline{n}) + \{1 + vE(\underline{r})\}^2\,E(\underline{n}^2). \tag{6.2.32}$$

Then, using (6.2.18) and $\varrho = \lambda E(\underline{h})$,

$$E(\underline{w}) = \frac{\lambda[vE(\underline{r}^2)\,E(\underline{n}) + \{1 + vE(\underline{r})\}^2 E(\underline{n}^2)]}{2[1 - \lambda\{1 + vE(\underline{r})\}\,E(\underline{n})]}. \tag{6.2.33}$$

This then is the average job waiting-time under general conditions for the case repair-resume. In the former case one had: $\underline{n} = 1$ and $\underline{r} = T$. Then $E(\underline{n}) = E(\underline{n}^2) = 1$, $E(\underline{r}) = T$, $E(\underline{r}^2) = T^2$. With those values (6.2.33) is consistent with the former result (6.2.23). For a full account of the use of combinatorial Laplace Transform methods in priority theory cf. Conway *et al.* (1967), which also gives an excellent bibliography.

After completion of the manuscript the author realized that there is a discipline that is strongly related to the present chapter, the *Theory of Reliability*. More information could possibly be obtained by consulting Barlow and Proschan (1965).

7. RESTRICTED AVAILABILITY

7.1. Introduction

Whatever the dimensioning criterion for service systems may be—whether a subjective measure like probability of blocking or average waiting-time, or a more objective econometric measure—it always turns out that the larger the arrival rates the larger the overall server efficiency η (cf. Fig. 2.2). This observation is a strong plea for concentration of traffic. This invariably leads to conflicts with other requirements, that have to be dealt with by compromise.

Consider, for example, an area with a number of local telephone exchanges $L_1, \ldots, L_m$. It is possible to interconnect those exchanges by a mesh-network: groups of *direct lines* are provided between any pair of exchanges. On the one hand, traffic is routed in this case via the shortest ways. On the other hand, many of those groups of direct lines will carry very little traffic and hence have poor efficiencies. Another possibility is a *star-network*: all exchanges are connected to a *central exchange* C by two groups of lines, one for each direction. Now, a connection $L_i \rightarrow L_j$ follows the route $L_i \rightarrow C \rightarrow L_j$. Per connection more mileage of cable is used. But the groups of lines now are fewer, they each carry more traffic and hence have a better efficiency than in the mesh-network.

Standard practice is a compromise. There is a star-network supplemented by groups of direct lines between exchanges that have a strong relationship (e.g. neighbouring cities). A demand for a connection $L_i \rightarrow L_j$ preferably uses a free item of the direct lines $L_i \rightarrow L_j$, if provided. When no direct line is free the *alternate route* $L_i \rightarrow C \rightarrow L_j$ is used. The traffic following this route is called the *overflow traffic*. All overflow traffics issuing from L_i share the use of the group of lines $L_i \rightarrow C$. Hence we shall call them *common lines*. As the direct lines now "skim" the traffic, they need not be so numerous. Hence, they may have better efficiencies than in the pure mesh-network case. The relatively rare overflowing demands issuing from L_i (and destined to various exchanges) now are grouped together and routed via $L_i \rightarrow C$. This group of common lines also handles the total traffics to exchanges L_j that are not directly attainable from L_i. The concentration of all those traffics on the group of lines $L_i \rightarrow C$ makes for a good efficiency for those lines too.

Now, the question of blocking may be raised: what percentage of the traffic $L_i \rightarrow L_j$ finds both the group of direct lines $L_i \rightarrow L_j$ (if provided) engaged as well as the group of common lines $L_i \rightarrow C$? Let us assume that the local exchange L_i produces traffic flows to the other exchanges $L_1, \ldots, L_m$ (L_i excluded) with arrival rates $\varrho_1, \ldots, \varrho_m$. Those traffics dispose of $c_1, \ldots, c_m$ direct lines. Moreover, there are c common lines $L_i \rightarrow C$ to accommodate all overflow traffic (cf. Fig. 7.1, where a square denotes a line). Demands from the traffic source group ($\rightarrow L_j$) can seize a free common line only in case all c_j direct lines $L_i \rightarrow L_j$ are occupied. There is supposed to be no waiting facility. In this connecting scheme we no longer have *full availability*, as the direct lines are not available to all sources, but only to part of them. The scheme is said to have *restricted availability*.

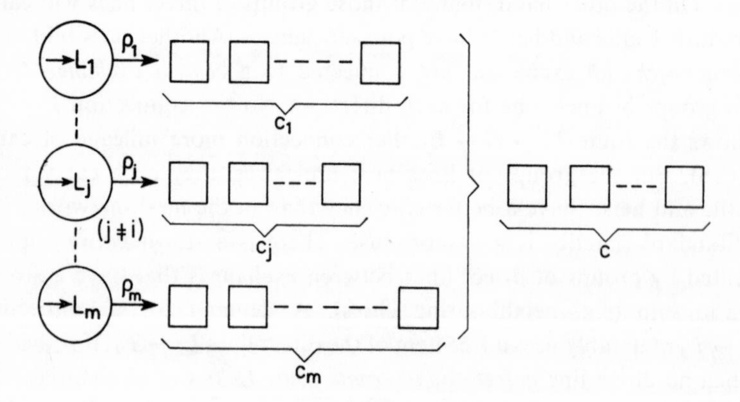

FIG. 7.1. Routing of overflow traffic.

Another cause for the occurrence of restricted availability may be the following. When a traffic flow is very large this may call for a very large number of servers. Now, there may be technical impediments to make all c servers available to all sources. In many instances a source cannot obtain access to more than k ($<c$) servers. The availability is restricted to k (e.g. 10 or 20 in telephony). It then is standard practice to split up the group of sources into a number of subgroups that each have their own k *choices* among the c servers. Demands from such a group first try the first choice device; if this is busy they try the second choice device, etc. If all k choices are busy, the demand is lost. Owing to this *hunting order*, higher-order choices will "catch" less traffic than lower-order ones. In view of this fact parallelization of higher-order choices

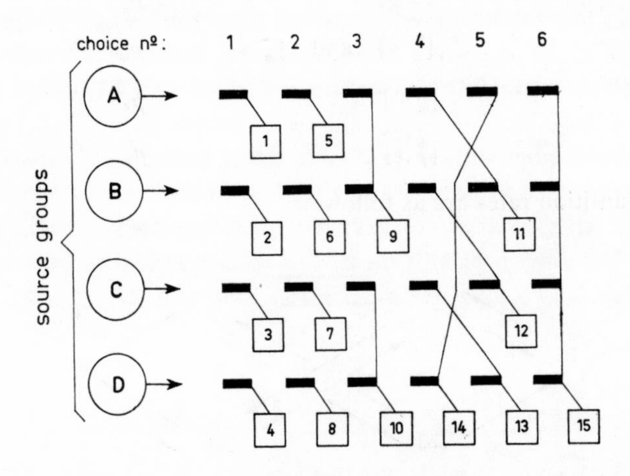

FIG. 7.2. Example of grading.

is applied which increases the efficiency of the servers appearing in those higher-order choices. In this way *gradings* come into existence. Figure 1.1 is a very simple one. Figure 7.2 shows a grading with so-called *individual*, *partly common* and *common servers*. Very often gradings are still more complicated. Again one can ask for the probabilities of blocking for the various subgroups of sources.

7.2. Restricted availability; exact evaluation of probabilities of blocking

We shall consider the elementary case of Fig. 1.1 under the simple assumption of exponential (1) holding-time. We denote unoccupied servers by dashes, occupied ones by asterisks. We describe a number of states and their steady-state probabilities. Configurations that are equiprobable on account of symmetry are taken together to form one state:

State	Description	Probability
[1]	$[^{-}_{-}\ -]$	p_1
[2]	$[^{*}_{-}\ -]$ and $[^{-}_{*}\ -]$	p_2
[3]	$[^{-}_{-}\ *]$	p_3
[4]	$[^{*}_{-}\ *]$ and $[^{-}_{*}\ *]$	p_4
[5]	$[^{*}_{*}\ -]$	p_5
[6]	$[^{*}_{*}\ *]$	p_6

The transition rates are as follows:

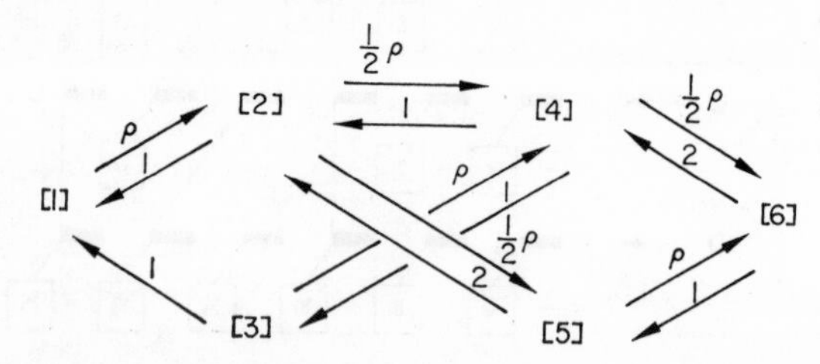

From this scheme the following birth-and-death equations result:

$$
\left.\begin{array}{lllllll}
0 = & -\varrho p_1 & + 1p_2 & + 1p_3 & & & \\
0 = & \varrho p_1 & - (\varrho + 1)\, p_2 & & + 1p_4 & + 2p_5 & \\
0 = & & & -(\varrho + 1)\, p_3 & + 1p_4 & & \\
0 = & & \tfrac{1}{2}\varrho p_2 & + \varrho p_3 & - (\tfrac{1}{2}\varrho + 2)\, p_4 & & + 2p_6 \\
0 = & & \tfrac{1}{2}\varrho p_2 & & & - (\varrho + 2)\, p_5 & + 1p_6 \\
0 = & & & & \tfrac{1}{2}\varrho p_4 & + \varrho p_5 & - 3p_6
\end{array}\right\} \quad (7.2.1)
$$

and

$$p_1 + \cdots + p_6 = 1. \quad (7.2.2)$$

Dropping one equation of (7.2.1) this system can be solved numerically. Finally the probability of blocking is

$$B = \tfrac{1}{2}p_4 + 1p_6 \tag{7.2.3}$$

as in the state 4 ($[{}^{*}_{-}\ *]$ and $[{}^{-}_{*}\ *]$) 50% of demands are lost, in state 6, 100%.

When the grading becomes more complicated or when the law of distribution of holding-times is different, the set of equations becomes rapidly unsolvable in practice. When there are c servers, when there is no symmetry and the holding-times have an Erlang-k distribution, the number of equations is $(k + 1)^c$, which "explodes" very quickly.

It is generally believed that the probabilities of blocking in gradings do not depend heavily on the law of distribution of holding-times. An authority like C. Palm (in early days) even believed in total independence (Palm, 1938). In order to check this hypothesis, the equivalent of the last problem has also been investigated under Erlang-2 assumption. Now, each server may be in one of three states: – (non-occupied), 1 (first phase of occupation) or 2 (second phase). In the case of symmetry pairs of equiprobable configurations may be taken together into one state. In the following scheme state 2 does not only consist of $[{}^{1}_{-}\ -]$, but also comprises $[{}^{-}_{1}\ -]$, etc. The states, etc., are

State	Description	Probability	State	Description	Probability
1	$[{}^{-}_{-}\ -]$	p_1	10	$[{}^{1}_{-}\ 2]$, $[{}^{-}_{1}\ 2]$	p_{10}
2	$[{}^{1}_{-}\ -]$, $[{}^{-}_{1}\ -]$	p_2	11	$[{}^{1}_{1}\ 1]$	p_{11}
3	$[{}^{-}_{-}\ 1]$	p_3	12	$[{}^{2}_{-}\ 2]$, $[{}^{-}_{2}\ 2]$	p_{12}
4	$[{}^{2}_{-}\ -]$, $[{}^{-}_{2}\ -]$	p_4	13	$[{}^{2}_{1}\ 1]$, $[{}^{1}_{2}\ 1]$	p_{13}
5	$[{}^{1}_{-}\ 1]$, $[{}^{-}_{1}\ 1]$	p_5	14	$[{}^{2}_{2}\ -]$	p_{14}
6	$[{}^{1}_{1}\ -]$	p_6	15	$[{}^{1}_{1}\ 2]$	p_{15}
7	$[{}^{-}_{-}\ 2]$	p_7	16	$[{}^{2}_{1}\ 2]$, $[{}^{1}_{2}\ 2]$	p_{16}
8	$[{}^{2}_{-}\ 1]$, $[{}^{-}_{2}\ 1]$	p_8	17	$[{}^{2}_{2}\ 1]$	p_{17}
9	$[{}^{2}_{1}\ -]$, $[{}^{1}_{2}\ -]$	p_9	18	$[{}^{2}_{2}\ 2]$	p_{18}

The transition-matrix is the following:

T=

i \ j	1	2	3	4	5	6	7	8	9	10	11	12	13	14	15	16	17	18
1	$-\rho$	ρ	·	·	·	·	·	·	·	·	·	·	·	·	·	·	·	·
2	·	$-\rho-2$	·	2	$\frac{1}{2}\rho$	$\frac{1}{2}\rho$	·	·	·	·	·	·	·	·	·	·	·	·
3	·	·	$-\rho-2$	·	ρ	·	2	·	·	·	·	·	·	·	·	·	·	·
4	2	·	·	$-\rho-2$	·	·	·	$\frac{1}{2}\rho$	$\frac{1}{2}\rho$	·	·	·	·	·	·	·	·	·
5	·	·	·	·	$-\frac{\rho}{2}-4$	·	·	2	·	2	$\frac{1}{2}\rho$	·	·	·	·	·	·	·
6	·	·	·	·	·	$-\rho-4$	·	·	4	·	ρ	·	·	·	·	·	·	·
7	2	·	·	·	·	·	$-\rho-2$	·	·	ρ	·	·	·	·	·	·	·	·
8	·	·	2	·	·	·	·	$-\frac{\rho}{2}-4$	·	·	·	2	$\frac{1}{2}\rho$	·	·	·	·	·
9	·	2	·	·	·	·	·	·	$-\rho-4$	·	·	·	ρ	2	·	·	·	·
10	·	2	·	·	·	·	·	·	·	$-\frac{\rho}{2}-4$	·	2	·	·	$\frac{1}{2}\rho$	·	·	·
11	·	·	·	·	·	·	·	·	·	·	-6	·	4	·	2	·	·	·
12	·	·	·	2	·	·	2	·	·	·	·	$-\frac{\rho}{2}-4$	·	·	·	$\frac{1}{2}\rho$	·	·
13	·	·	·	·	2	·	·	·	·	·	·	·	-6	·	·	2	2	·
14	·	·	·	4	·	·	·	·	·	·	·	·	·	$-\rho-4$	·	·	ρ	·
15	·	·	·	·	·	2	·	·	·	·	·	·	·	·	-6	4	·	·
16	·	·	·	·	·	·	·	·	2	2	·	·	·	·	·	-6	·	2
17	·	·	·	·	·	·	·	4	·	·	·	·	·	·	·	·	-6	2
18	·	·	·	·	·	·	·	·	·	·	·	4	·	2	·	·	·	-6

Then the system of equations

$$\sum_{i=1}^{18} p_i t_{ij} = 0 \quad \text{is solved together with} \quad \sum_{1}^{18} p_i = 1.$$

As in states 5, 8, 10 and 12, 50% of demands are lost and in states 11, 13, 15, 16, 17 and 18, 100%, the probability of blocking is

$$B = \tfrac{1}{2}(p_5 + p_8 + p_{10} + p_{12}) + p_{11} + p_{13} + p_{15} + p_{16} + p_{17} + p_{18}. \tag{7.2.4}$$

The results (7.2.3 and 7.2.4) have been calculated for various values of ϱ within the range $10^{-4}, \ldots, 50$. They always agreed in three significant figures. So these results do not disprove the general belief.

7.3. Approximate determination of probabilities of blocking; characteristics of overflow traffic

In the former section it was made clear that exact determination of probabilities of blocking in restricted availability cases is beyond our power except for the very simplest configurations. In view of the importance of those cases, however, approximate methods have been developed. Though we do not deal with approximate methods, approximations sometimes use theoretical results that fall within the scope of the present book. They will therefore be given some consideration.

Consider the case of alternate routing described in Section 7.1 (cf. Fig. 7.1). The traffic overflowing from source group $(\rightarrow L_j)$ has a density $\alpha_j = \varrho_j E_{1,c_j}(\varrho_j)$. Hence, the total traffic offered to the common lines is $\alpha = \sum_1^m \alpha_j$. Merging of traffic generally has a tendency towards a Poisson arrival process. Consequently, a first estimate for the total loss could be $\alpha E_{1,c}(\alpha)$. This loss then could be distributed over the several groups in proportion to the α_j. The losses found in this way generally are too low. This cannot be wondered at: we have totally neglected the "peakedness" of overflow traffic, i.e. the fact that overflowing demands arrive in "clusters".

R. I. Wilkinson (1956) has developed an ingenious method of dealing with this effect. It is called the *equivalent random method*. In the simple approximate method given above, the overflow traffics were described merely by their densities α_j, which characterization clearly is too crude. Now, Wilkinson characterizes overflow traffics by *two* well-chosen parameters. For the definition of those parameters the overflow traffic of the group $(\rightarrow L_j)$ is fictitiously offered to an infinite group of common lines (cf. Fig. 7.3). The numbers $\underline{r}_j$ and $\underline{s}_j$ of simultaneously occupied

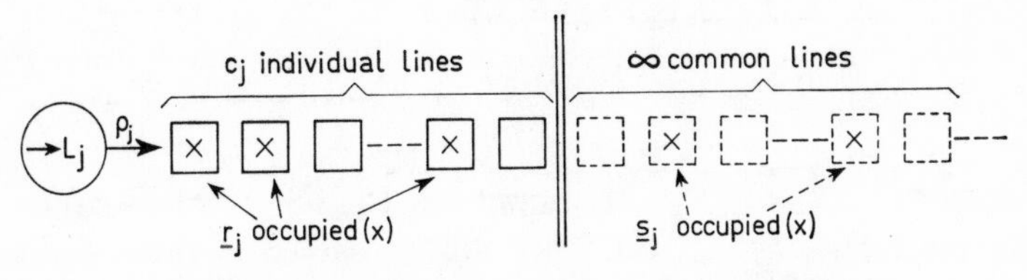

FIG. 7.3. Characterization of an overflow traffic.

direct and common lines (by demands stemming from the ($\to L_j$) traffic) are correlated stochastic variables. Now, the two parameters chosen by Wilkinson are the average and the variance of $\underline{s}_j$ when $\underline{r}_j$ is unknown. As there is no loss of traffic in the infinite group of common lines, the average of $\underline{s}_j$ must be $\alpha_j = \varrho_j E_{1,c_j}(\varrho_j)$. Let us now assume that $v_j := \operatorname{var}(\underline{s}_j)$ is also known. Let *all* overflow traffics be offered simultaneously to the group of common lines. The number $\underline{s}$ of simultaneous occupations in this group clearly is the sum of the number of occupations stemming from the different groups:

$$\underline{s} = \underline{s}_1 + \cdots + \underline{s}_m .$$

Now, $\underline{s}_1, \ldots, \underline{s}_m$ obviously are mutually independent. Hence, the average α and the variance v of $\underline{s}$ are:

$$\alpha := E(\underline{s}) = \sum_1^m \alpha_j, \qquad v := \operatorname{var}(\underline{s}) = \sum_1^m v_j. \tag{7.3.1}$$

So, if we consider α and v to be characteristic for the total composite overflow traffic, those quantities are simply obtained by addition of those quantities for the m constituents. Suppose there is *one* fictitious

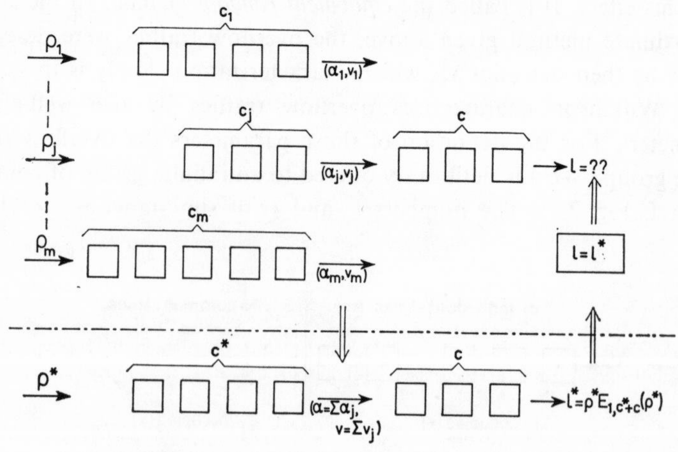

FIG. 7.4. Principle of the "equivalent random" method.

Poisson traffic with arrival rate ϱ^* that, after being skimmed by c^* lines, yields an overflow traffic with the *same* characteristics α and v. Then Wilkinson conjectures that *this* overflow traffic and the original composite overflow do not only behave in the same way when offered to an infinite group of common lines, but also when the number of common lines is finite (cf. Fig. 7.4). Hence the total lost traffic l of the original configuration is supposed to be equal to the loss l^* of the fictitious traffic after having been skimmed by $c^* + c$ lines in total, i.e. $l = l^* = \varrho^* E_{1,c^*+c}(\varrho^*)$. This loss then is distributed among the m groups proportional to $\alpha_1, \ldots, \alpha_m$. From the separate losses follow the m probabilities of blocking per group.

In order for the method to be workable one needs:

(i) a method for evaluating the average α and variance v of the overflow traffic of one group as a function of the arrival rate ϱ and the number of direct lines c (the group index $_j$ has been dropped);

(ii) tables or graphs of those functions.

These requirements are met by Wilkinson. Moreover, the method has been experimentally verified.

Now, we consider the case of Fig. 7.3, dropping the suffixes j. Let f_{rs} be the steady-state probability of the simultaneous occupation of r direct and s common lines. The birth-and-death equations are:

$$0 = -(r + s + \varrho) f_{rs} + (r + 1) f_{r+1,s} + (s + 1) f_{r,s+1} + \varrho f_{r-1,s} \qquad (r < c;\ s = 0, 1, \ldots), \qquad (7.3.2)$$

$$0 = -(c + s + \varrho) f_{cs} + \varrho f_{c,s-1} + (s + 1) f_{c,s+1} + \varrho f_{c-1,s} \qquad (s = 0, 1, \ldots) \qquad (7.3.3)$$

and

$$\sum_{r=0}^{c} \sum_{s=0}^{\infty} f_{rs} = 1. \qquad (7.3.4)$$

Assume (7.3.2) to be valid for $r \geqq c$ too (defining fictitious quantities $f_{c+1,s}, \ldots$). Let the following single and double generating functions be introduced:

$$f_{rs} \triangleq F_r(y) \triangleq F(x, y). \qquad (7.3.5)$$

Then the equations given above yield

$$(1 - x)\frac{\partial F}{\partial x} + (1 - y)\frac{\partial F}{\partial y} = \varrho(1 - x) F, \qquad (7.3.6)$$

$$0 = (-c - \varrho + \varrho x) F_c + (1 - y)\frac{\partial F_c}{\partial y} + \varrho F_{c-1}, \qquad (7.3.7)$$

$$F_c^1(1) = 1. \qquad (7.3.8)$$

The general solution of (7.3.6) is

$$F(x, y) = K\left(\frac{1 - y}{1 - x}\right) e^{-\varrho(1-x)}, \qquad (7.3.9)$$

where K is an arbitrary function. It is assumed that K is expandable in a power series:

$$K\left(\frac{1 - y}{1 - x}\right) = \sum_{i=0}^{\infty} \beta_i \left(\frac{1 - y}{1 - x}\right)^i. \qquad (7.3.10)$$

Taking the arithmetic function generated by (7.3.9) yields (cf. A 7′):

$$F_r(y) = \sum_{i=0}^{\infty} \beta_i \varphi_r^i (1 - y)^i. \qquad (7.3.11)$$

Comparison of (7.3.2) and (7.3.3) yields $(c + 1) f_{c+1,s} = \varrho f_{c,s-1}$ ($s = 0, 1, \ldots$), which is equivalent to

$$(c + 1) F_{c+1}(y) = \varrho y F_c(y). \qquad (7.3.12)$$

When the series (7.3.11) is inserted here and the coefficients of $(1 - y)^i$ in the resulting equation are equated to zero, we obtain

$$(c + 1) \beta_i \varphi_{c+1}^i = \varrho \beta_i \varphi_c^i - \varrho \beta_{i-1} \varphi_c^{i-1}, \qquad (7.3.13)$$

or, using (A 18),

$$i \beta_i \varphi_c^{i+1} = -\varrho \beta_{i-1} \varphi_c^{i-1} \quad (i = 1, 2, \ldots). \qquad (7.3.14)$$

Multiple application of this relation yields

$$\beta_i = \frac{(-\varrho)^i}{i!} \frac{\varphi_c^1 \varphi_c}{\varphi_c^{i+1} \varphi_c^i} \beta_0. \qquad (7.3.15)$$

From (7.3.8 and 7.3.11) it follows that $\beta_0 = 1/\varphi_c^1$. Now, all unknown quantities have been determined. We obtain in particular

$$\begin{aligned} \beta_1 &= -\varrho\varphi_c/\varphi_c^1\varphi_c^2, \\ \beta_2 &= \tfrac{1}{2}\varrho^2\varphi_c/\varphi_c^2\varphi_c^3. \end{aligned} \tag{7.3.16}$$

The generating function $H(y)$ of the probability h_s of $\underline{s} = s$ occupied common lines, irrespective of the number of occupied direct lines, is

$$h_s \triangleq H(y) = F_c^1(y) = \sum_{i=0}^{\infty} \beta_i \varphi_c^{i+1}(1-y)^i. \tag{7.3.17}$$

The average and the second factorial moment are

$$\begin{aligned} \alpha = E(\underline{s}) &= H'(1) = -\beta_1\varphi_c^2 = \varrho\varphi_c/\varphi_c^1 = \varrho E_{1,c}(\varrho), \\ E\{\underline{s}(\underline{s}-1)\} &= H''(1) = 2\beta_2\varphi_c^3 = \varrho^2\varphi_c/\varphi_c^2. \end{aligned} \tag{7.3.18}$$

Hence, the variance v is

$$v = \mathrm{var}\,(\underline{s}) = E\{\underline{s}(\underline{s}-1)\} + E(\underline{s}) - E^2(\underline{s}) = \varrho^2\varphi_c/\varphi_c^2 + \alpha - \alpha^2. \tag{7.3.19}$$

The result (7.3.18) is the known average. The derivation of v given above follows a suggestion made by J. Riordan in appendix I to Wilkinson's paper (1956, p. 507) in connection with Kosten (1937, 1942). In Wilkinson–Riordan a similar derivation may be found.

8. ARRIVAL AND SERVICE IN BATCHES

8.1. Introduction

Up to now we always have assumed that demands are made individually and that the server serves those demands one at a time. There are, however, cases where those conditions are not met. In the booking process of airlines passengers may arrive singly or in clusters. Elevators handle passengers in batches. We shall pay some attention here to those alternatives. No cases will be considered, however, with both arrival and servicing in batches. It is assumed that there is one server only, with waiting facility.

In Section 8.2 we shall consider the case of random arrivals (Poisson process) of batches of stochastically variable size. The holding-time is assumed to be exponential (1). The average waiting-time will be determined.

In Sections 8.3 and 8.4 it is assumed that demands are served in batches, which may be of variable size. The holding-time for those batches will generally be of variable duration. Though it is possible to consider an arbitrary distribution of holding-times, it is *not* easy to cope with the main cause of this variability, viz. the dependence on the number of items in the batch. For this reason the constant holding-time case only will be dealt with.

In Section 8.3 the server, upon being released, starts processing a new batch of waiting demands not exceeding the batch capacity c. If there are more than c demands waiting, the remaining ones stay waiting. If no demand is waiting when the server is released, the latter waits for the first arrival. In Section 8.4 there is no server capacity restriction. In this case, however, the server does not resume its bulk service before there are at least b (>1) waiting demands (the "custodian's problem"). In Section 8.5 a variant of the latter problem will be discussed.

There is a large literature on the subject of arrival and service in batches. Apart from some publications cited in the next sections, one may consult more ample references in Saaty (1961) chapter 7, and Cohen (1969) III.2.

In many cases analysis is not possible or leads to very unwieldy results. Often the difficulty can be overcome by switching to numerical methods at an appropriate point (cf. the examples of Sections 8.3 and 8.5).

8.2. Arrivals in batches; individual service (exponential holding-time)

Batches of demands arrive according to a Poisson process. The number of demands per batch also is a stochastic variable. Numbers of items in successive batches are uncorrelated. This model of arrivals—sometimes called a *compound Poisson process*—can also be described as follows. There are Poisson arrival processes for singletons, pairs, etc., which arrivals together form our arrival process. Let λ_i ($i = 1, 2, \ldots$) be the arrival rate of singletons, pairs, etc. The total arrival rate of batches will be denoted by $\lambda := \sum_1^\infty \lambda_i$. The average number of demands per unit of time will be $\varrho := \sum_1^\infty i\lambda_i$. There is one server with exponential (1) holding-times. There is an infinite waiting facility. It is required to determine the average waiting-time.

Let $[r]$ be the state with $\underline{r} = r$ items in the system ($r = 0, 1, \ldots$) and p_r its steady-state probability. The state $[r]$ can change into $[r + i]$ ($i = 1, 2, \ldots$) at transition rate λ_i and into $[r - 1]$ at rate 1 (if $r > 0$ only). The birth-and-death equations are:

$$0 = \sum_{i=1}^{r} \lambda_i p_{r-i} - \{\lambda p_r + (p_r - \delta_r^0 p_0)\} + p_{r+1} \quad (r = 0, 1, 2, \ldots), \qquad (8.2.1)$$

where

$$\sum_0^\infty p_r = 1. \qquad (8.2.2)$$

Now, the following generating functions are introduced (with $\lambda_0 := 0$):

$$p_r \triangleq P(x),$$
$$\lambda_r \triangleq \Lambda(x) = \sum_0^\infty \lambda_r x^r, \tag{8.2.3}$$

and hence

$$\Lambda(1) = \lambda, \quad \Lambda'(1) = \varrho, \quad \Lambda''(1) = \sum_{i=2}^\infty i(i-1)\lambda_i. \tag{8.2.4}$$

Then (8.2.1) yields (cf. A 5)

$$0 = \Lambda(x)P(x) - (\lambda + 1)P(x) + P(0) + \frac{P(x) - P(0)}{x},$$

or

$$P(x) = (1-x)P(0)/[1 - x\{\lambda + 1 - \Lambda(x)\}] \tag{8.2.5}$$

and (8.2.2):

$$P(1) = 1. \tag{8.2.6}$$

With the use of (8.2.4) one obtains

$$P(x) = \frac{P(0)}{1-\varrho}\left[1 - \frac{1-x}{1-\varrho}\{\tfrac{1}{2}\Lambda''(1) + \varrho\} + O\{(1-x)^2\}\right]. \tag{8.2.7}$$

Comparison with (8.2.6) shows that $P(0) = 1 - \varrho$, which is the value we expect. Owing to the fact that arrivals are in clusters, it is not the complement of the probability of delay. As a rule all i demands of an arriving batch of i items incur delay. If the batch arrives when the system is in [0], however, one of the i items is served immediately. Hence, the average number of demands that are delayed per unit of time equals:

$$(1-\varrho)\sum_1^\infty (i-1)\lambda_i + \varrho\sum_1^\infty i\lambda_i = \varrho - (1-\varrho)\lambda.$$

The probability of delay D is obtained by dividing this quantity by the demand arrival rate ϱ: $D = 1 - (1-\varrho)\lambda/\varrho$. In the case of mere singletons it reduces to $\varrho = \lambda = D$.

From (8.2.7 and 8.2.4) it results that

$$E(\underline{r}) = P'(1) = \{\tfrac{1}{2}\sum i(i-1)\lambda_i + \varrho\}/(1-\varrho). \tag{8.2.8}$$

The average number in the queue is obtained by subtracting the occupancy ϱ of the server. Then dividing the result by the demand arrival rate ϱ the average waiting-time per demand is obtained:

$$w = \left\{\varrho + \frac{1}{2\varrho} \sum_{2}^{\infty} i(i-1)\,\lambda_i\right\} \Big/ (1-\varrho). \tag{8.2.9}$$

When dividing this by D, the average waiting-time W per delayed demand may be had.

8.3. Service in batches of capacity c; constant holding-time

The demands arrive individually according to a Poisson process with density λ. The server is able to handle collectively any number of demands not in excess of c. It needs a constant holding-time $\underline{h} = 1$ for such a batch. If no demand is waiting when the server becomes available, the latter will help the first arriving demand.

It would be perfectly possible to use the method of birth-and-death equations. The more combinatorial approach used here is a bit more straightforward.

The maximum number of demands that could possibly be handled is c per unit of time. So for $\lambda > c$ no stationarity is possible. On the other hand, it is intuitively clear that for $\lambda < c$ stationarity may obtain.

When we look at the server, we may distinguish periods of unit length during which the server helps a batch and periods where the server is idle. We shall call them *service periods* and *idle periods*. Consecutive service periods may either be linked together or separated by one idle period.

Let f_r and g_r be the probability of $\underline{r} = r$ waiting demands at the beginning and at the end of a service period, respectively.

During a service period the number of waiting demands increases by a random variable with distribution $\varphi_r = e^{-\lambda}\lambda^r/r!$ which is not dependent on the number of waiting demands at the beginning. As any combination of $r - i$ original waiting demands and i arrivals during the service period leaves r waiting demands at the end we have

$$g_r = \sum_{i=0}^{r} \varphi_i f_{r-i} \quad (r = 0, 1, \ldots). \tag{8.3.1}$$

Between the end of one service period and the beginning of the next the number of waiting demands decreases by c or by so much less as is necessary to yield a non-negative result. Hence,

$$f_r = g_{r+c} + \delta_r^0 \sum_{i=0}^{c-1} g_i \quad (r = 0, 1, \ldots). \tag{8.3.2}$$

Together with $\sum_0^\infty f_r \, (= \sum_0^\infty g_r) = 1$ those sets of equations should be sufficient to determine both distributions.

Let us now assume that f_r and g_r have been obtained. A fraction g_0 of active periods is followed by an idle period. The duration of the latter is exponential $(1/\lambda)$. Consequently, the server is active during a fraction $\lambda/(\lambda + g_0)$ of total time and idle during a fraction $g_0/(\lambda + g_0)$. The first of those expressions equals the probability of delay:

$$D = \lambda/(\lambda + g_0). \tag{8.3.3}$$

The average number of waiting demands at the beginning of a service period is $\sum_1^\infty rf_r$. During this period it increases linearly by λ. Hence, the average total waiting-time during a service period is $\sum_1^\infty rf_r + \frac{1}{2}\lambda$. During idle periods there is no waiting. There is just one demand per idle period, which ends this period; it does not incur delay. Hence, the average total waiting-time per service period is incurred by a number of demands arising during one service period, i.e. λ on the average. Thus the average waiting-time per delayed demand is

$$W = \tfrac{1}{2} + \frac{1}{\lambda} \sum_1^\infty rf_r. \tag{8.3.4}$$

The method developed so far still requires the determination of f_r and g_r from (8.3.1 and 8.3.2) and $\sum f_r = 1$. We shall first discuss an analytical method that reflects existing literature for the arbitrary holding-time distribution case (Bailey, 1954; Downton, 1955; cf. also Saaty, 1961). When we introduce the generating functions:

$$f_r \triangleq F(x), \quad g_r \triangleq G(x) \tag{8.3.5}$$

and use $\varphi_r \triangleq e^{-\lambda(1-x)}$, equations (8.3.1 and 8.3.2) yield

$$F(x) = \frac{G(x) - \sum_{i=0}^{c-1} g_i x^i}{x^c} + \sum_{i=0}^{c-1} g_i, \tag{8.3.6}$$

and

$$G(x) = e^{-\lambda(1-x)} F(x), \tag{8.3.7}$$

and the norming equation is

$$F(1) = 1. \tag{8.3.8}$$

The generating function for g_{r+c} in (8.3.6) has been obtained by multiple application of (A 3). Substitution of (8.3.7) in (8.3.6) yields

$$F(x) = \sum_{i=0}^{c-1} (x^c - x^i)\, g_i / \{x^c - e^{-\lambda(1-x)}\}. \tag{8.3.9}$$

By the theorem of Rouché (cf. Copson, 1935) it may be shown that for $0 < \lambda < c$ the denominator possesses $c - 1$ roots $x_1, \ldots, x_{c-1}$ within the unit circle $|x| = 1$. Now, as $\sum f_r$ must exist, $F(x)$ is analytic within this circle. Hence, for the same values $x = x_1, \ldots, x_{c-1}$ that cause the denominator to vanish, the numerator in (8.3.9) must vanish as well:

$$\sum_{i=0}^{c-1} (x_j^c - x_j^i)\, g_i = 0 \quad (j = 1, \ldots, c - 1). \tag{8.3.10}$$

This then is a set of equations which determine g_i $(i = 0, \ldots, c - 1)$ up to a constant factor. Then (8.3.9) determines $F(x)$ apart from this factor, and that may be obtained from (8.3.8). Then (8.3.4) yields

$$W = \tfrac{1}{2} + \frac{1}{\lambda} F'(1). \tag{8.3.11}$$

The foregoing analysis is rather intractable when c is not very small. It contains the determination of a large set of zeros (mostly complex) $x_1, \ldots, x_{c-1}$. A better way is to switch over to a numerical method in an earlier stage. The derivation of (8.3.1) and (8.3.2) does *not* necessarily suppose stationarity to exist. The first equation relates the distributions of the number of waiting demands at the beginning and end of *any* service period; the second equation relates the distributions at the beginning of *any* service period to that existing at the end of the aforegoing service period. Now, let f_{mr} and g_{mr} be the distributions of waiting demands at the beginning and end, respectively, of the mth service period ($m = 0, 1, 2, \ldots$). Here $m = 0$ denotes some service period, arbitrarily taken to be the zeroth. Then the equivalents of (8.3.1 and 8.3.2) are

$$g_{mr} = \sum_{i=0}^{r} \varphi_i f_{m,r-i}, \tag{8.3.12}$$

$$f_{m+1,r} = g_{m,r+c} + \delta_r^0 \sum_{i=0}^{c-1} g_{mi} \quad (m, r = 0, 1, \ldots). \tag{8.3.13}$$

If, for example, we start with $f_{or} = \delta_r^0$, the alternate application of those equations yields the distributions $g_{0r}, f_{1r}, g_{1r}, \ldots$. Supposing $\lambda < c$, a steady state will gradually develop:

$$\left.\begin{aligned} f_r &= \lim_{m\to\infty} f_{mr}, \\ g_r &= \lim_{m\to\infty} g_{mr}. \end{aligned}\right\} \tag{8.3.14}$$

Then (8.3.3) and (8.3.4) yield the probability of delay D and the average waiting time for delayed demands. There are no specific numerical difficulties. The number of variables is delimited by $r = 0, 1, \ldots, R$, where R is some large number (such that the f_R and g_R will be nearly zero). The values g_{mr} ($r > R$) occurring in (8.3.13) are taken to be identically zero. The algorithm of using alternatingly (8.3.12) and (8.3.13) is ended as soon as f_{mr} and g_{mr} remain constant in, say, four decimals for all r. It is possible that rounding errors cumulate to such an extent, that $\sum f_r$ and $\sum g_r$ ultimately will deviate substantially from 1. A final norming may then be necessary. After the determination of f_r and g_r ($r = 0, \ldots, R$) D and W result from (8.3.3) and (8.3.4) (where the summing is to be stopped at $r = R$). Figure 8.1 shows W as a function of λ for various values of c. The number of iterations necessary increases with the value of c and λ. It varied from 3 to 100.

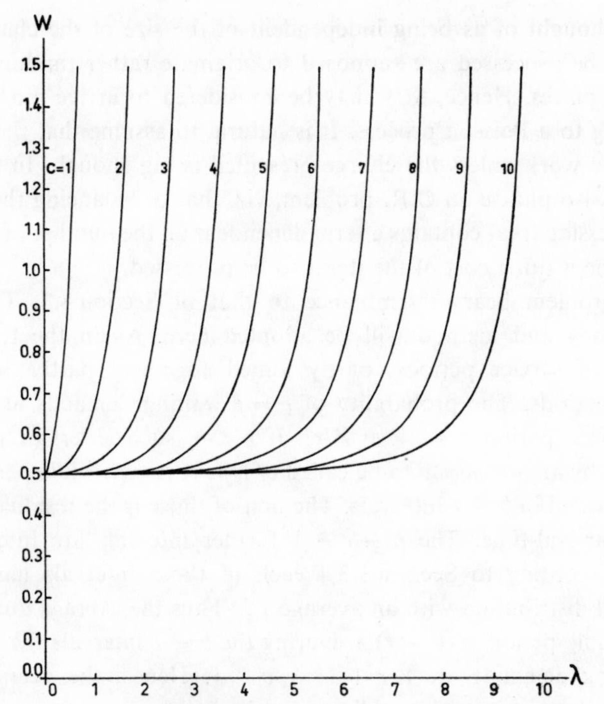

FIG. 8.1. Average waiting-time per delayed demand; helping in batch.

8.4. The "custodian's problem"

In a museum the custodian is organizing guided tours with a constant duration of one unit of time. Visitors arrive according to a Poisson process with density λ. They have to wait until the custodian begins a new tour. At this tour the custodian takes along all waiting visitors. After completion of a tour the custodian starts a new tour if at least b prospective visitors are waiting. If not so, he waits until the number of b visitors is obtained, at which moment he immediately begins a new tour. It is primarily required to determine the average waiting-time w per visitor.

There are technical situations to which the foregoing model seems to apply. Consider an enamelling oven with a large capacity. The processing time is thought of as being independent of the size of the charge. The items to be processed are supposed to originate rather randomly from different places. Hence, they may be considered to arrive more or less according to a Poisson process. It is natural to assume that the oven is not set to work unless the charge presented is big enough. In this case one can also phrase an O.R. problem, viz. that of balancing the cost of the processing (that contains a term dependent on the number of charges) and of the waiting cost of the items to be processed.

The problem bears resemblance to that of Section 8.3. The same terminology and method will be adopted here. Again the total time consists of service periods, partly joined together, partly separated by idle periods. The probability of $\underline{r} = r$ waiting demands at the end of a service period is $\varphi_r = e^{-\lambda}\lambda^r/r!$. If $r < b$, such a service period is followed by an *idle period* to be called *of type r*. A type r idle period may be subdivided in $b - r$ intervals. The first of those is the residual part of an interarrival-time. The $b - r - 1$ further intervals are interarrival-times. According to Section 1.3.4 each of those intervals has an exponential distribution with an average $1/\lambda$. Thus the average duration of a type r idle period is $(b - r)/\lambda$. During the $b - r$ intervals the numbers of waiting demands are $r, r + 1, \ldots, b - 1$. Hence, the average total waiting-time during a type r idle period is

$$\{r + (r + 1) + \cdots + (b - 1)\} \cdot 1/\lambda = (b + r - 1)(b - r)/2\lambda .$$

During service periods the average total waiting-time is equal to $\lambda/2$.

Now, consider a service period together with the immediately following idle period, if any. This combination will be called a *gross period*. The average duration T of the gross period is

$$T = 1 \cdot 1 + \sum_{r=0}^{b-1} \varphi_r \cdot (b - r)/\lambda, \tag{8.4.1}$$

the inverse of which equals the probability of delay:

$$D = \frac{\lambda}{\lambda + \sum_{r=0}^{b-1} (b - r)\, \varphi_r} . \tag{8.4.2}$$

The average total waiting-time per gross period equals

$$E(\underline{w}_{\text{tot}}) = 1 \cdot \lambda/2 + \sum_{r=0}^{b-1} \varphi_r \cdot (b+r-1)(b-r)/2\lambda. \qquad (8.4.3)$$

As λT is the average number of demands per gross period, the average waiting-time per demand (= visitor) is equal to

$$w = \frac{\lambda + \dfrac{1}{\lambda}\sum_{r=0}^{b-1}(b+r-1)(b-r)\varphi_r}{2\left\{\lambda + \sum_{r=0}^{b-1}(b-r)\varphi_r\right\}}. \qquad (8.4.4)$$

By the use of generating functions the results (8.4.2 and 8.4.4) may be brought into the forms:

$$D = \frac{\lambda}{\lambda + \varphi_{b-1}^2}, \qquad (8.4.2')$$

$$w = \frac{\frac{1}{2}\lambda^2 + \lambda\varphi_{b-2}^2 + \varphi_{b-2}^3}{\lambda^2 + \lambda\varphi_{b-1}^2}. \qquad (8.4.4')$$

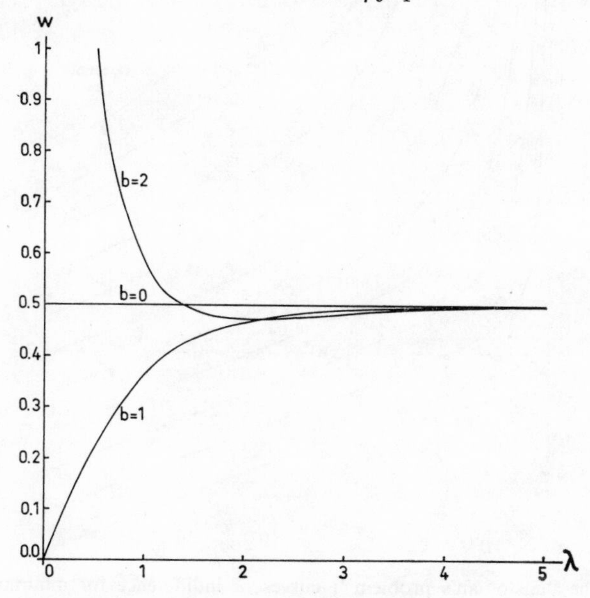

FIG. 8.2. The "custodian's problem"; average waiting-time.

In Fig. 8.2 w has been drawn as a function of λ for $b = 0, 1, 2$. When the objective is to minimize the average waiting-time one should always take $b = 1$. The small improvement possible by taking $b = 2$ for $\lambda > 2$ does not justify taking the risk of the large average waiting-time which would occur should λ happen to decrease.

Now, let k_1 be the cost of one service period (= tour) and k_2 the cost of one demand (= visitor) waiting for one unit of time. Let $p = k_1/k_2$ be the cost ratio. Per unit of time the expected number of service periods is D, the expected total waiting-time λw. Hence, minimization of total cost comes to

$$\operatorname*{Min}_{(b)} \frac{\lambda p + \frac{1}{2}\lambda^2 + \lambda \varphi^{2}_{b-2} + \varphi^{3}_{b-2}}{\lambda + \varphi^{2}_{b-1}}. \tag{8.4.5}$$

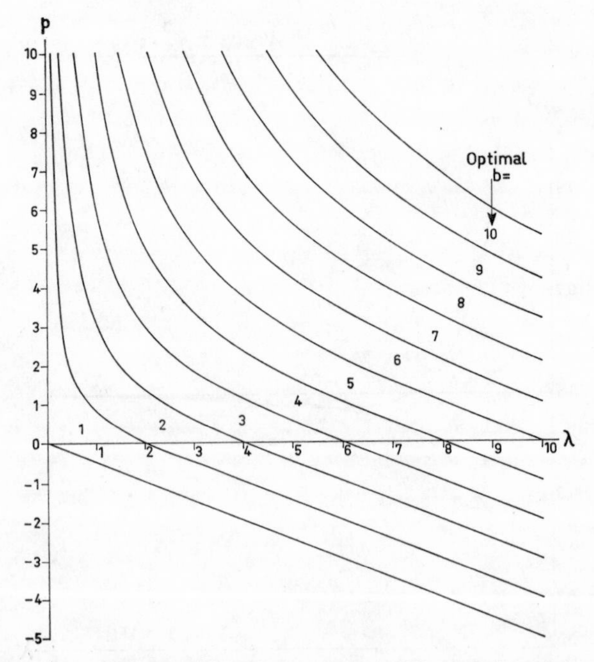

FIG. 8.3. The "custodian's problem"; curves of indifference for minimum number of tour participants.

In the p–λ plane the "curve of indifference $b/b+1$" is determined by

$$\frac{\lambda p + \frac{1}{2}\lambda^2 + \lambda\varphi_{b-2}^2 + \varphi_{b-2}^3}{\lambda + \varphi_{b-1}^2} = \frac{\lambda p + \frac{1}{2}\lambda^2 + \lambda\varphi_{b-1}^2 + \varphi_{b-1}^3}{\lambda + \varphi_b^2}. \quad (8.4.6)$$

By some tedious algebra (Kosten, 1967) this curve may be shown to be given by

$$p = b - \frac{\lambda}{2} + \frac{\varphi_{b-1}^3}{\lambda}. \quad (8.4.7)$$

The results are shown in Fig. 8.3.

8.5. The "unscheduled ferry problem"

A ferry transports cars between a port A and a port B. In both ports cars arrive according to Poisson processes with densities λ. The ferry needs one unit of time for a trip $A \to B$ or $B \to A$. Loading and unloading are supposed to take no time. The ferry does not sail according to a time table. It starts whenever the number of cars awaiting transport in the sailing direction is at least b. The capacity of the vessel is so large that on sailing it can take along all cars waiting in the port of departure. It is required to determine the average waiting-time for the cars.

This problem strongly resembles the last ones. It will be dealt with in a similar manner. The successive trips (in alternating directions) will be numbered $m = 0, 1, 2, \ldots$. In order to avoid circumstantial descriptions we shall assume that trip m goes from A to B and hence trip $m+1$ from B to A. By m^- and m^+ we shall denote the instants during trip m just after leaving A and just before reaching B. The instants $(m+1)^-$ and $(m+1)^+$ are defined correspondingly. Successive trips may be immediately following each other or show a gap, the so-called idle period, when the vessel is waiting for cars to complete the number b. The combination of a trip and the idle period following it, if any, will again be termed the gross period. By $[r, s]$ we denote the state of the car-queues: $\underline{r} = r$ items waiting in the port from which the ferry sailed for the last time, $\underline{s} = s$ cars waiting in the other port (i.e. the port of destination when the ferry is sailing and the port where the ship lies otherwise).

At m^-, $\underline{r} \equiv 0$, since the ferry took along all cars in A. At m^+, $\underline{r}$ is distributed according to a Poisson distribution: $\varphi_r = e^{-\lambda}\lambda^r/r!$. Moreover, at m^+, $\underline{r}$ and $\underline{s}$ are uncorrelated.

Let p_{ms} be the probability of $[0, s]$ at m^-, $\varphi_r \cdot q_{ms}$ the probability of $[r, s]$ at m^+.

The state $[r, s]$ at m^+ can arise from any $[0, i]$ state ($i = 0, \ldots, s$) at m^- by r arrivals in A and $s - i$ in B (probabilities φ_r and φ_{s-i}, respectively). Hence,

$$q_{ms} = \sum_{i=0}^{s} \varphi_{s-i} p_{mi}. \tag{8.5.1}$$

Now, consider how $[0, s]$ at $(m + 1)^-$ can originate from states at m^+. *Between m^+ and $(m + 1)^-$ A and B change roles!* At m^+, A is the port of departure, at $(m + 1)^-$ it is B. The state $[0, s]$ at $(m + 1)^-$ can arise in the following ways from states at m^+:

(i) from $[s, i]$ ($i \geqq b$), with certainty;

(ii) from $[s - j, b - i]$ ($j = 0, \ldots, s$; $i = 1, \ldots, b$), with a probability f_{ij} (defined as the probability that between the ferry's arrival in B and the subsequent ith car arrival in B the number of car arrivals in A is j).

Hence,

$$p_{m+1,s} = \varphi_s \sum_{i=b}^{\infty} q_{mi} + \sum_{i=1}^{b-1} q_{m,b-i} \sum_{j=0}^{s} f_{ij}\, \varphi_{s-j}. \tag{8.5.2}$$

Now, the probabilities f_{ij} must be determined. Under the conditions specified under (ii) the ferry's idle period in B consists of i exponential intervals, so that the duration of this idle period has an Erlang-i distribution. Let the average duration of this idle period temporarily be taken as a unit of time. Then both car arrival rates are i. The probability that the idle period lasts for $u(+du)$ is (cf. 4.4.1): $e^{-iu}i^i u^{i-1}du/(i - 1)!$. The conditional probability for j arrivals in A during this period then is $e^{-iu}(iu)^j/j!$. Hence, the probability for j car arrivals in A during the idle period is

$$f_{ij} = \int_0^{\infty} \frac{e^{-iu}i^i u^{i-1}}{(i-1)!} \cdot \frac{e^{-iu}(iu)^j}{j!}\, du,$$

or

$$f_{ij} = \binom{i+j-1}{j} 2^{-i-j}. \tag{8.5.3}$$

When the values of f_{ij} have been inserted in (8.5.2), the alternate use of (8.5.1) and (8.5.2) enables us to calculate $q_{0s}, p_{1s}, q_{1s}, \ldots$, starting arbitrarily with, for example, $p_{0s} = \delta_s^0$. It is intuitively clear that this will tend to steady-state distributions:

$$\left.\begin{aligned} p_s &:= \lim_{m\to\infty} p_{ms}, \\ q_s &:= \lim_{m\to\infty} q_{ms}. \end{aligned}\right\} \tag{8.5.4}$$

After these distributions have been obtained to a sufficient degree of stability (say in four significant figures) they may be used to evaluate the average waiting-time. In order to do so we shall determine the average duration of gross periods and the average total waiting-time per gross period.

Per gross period there is a probability $\varphi_k q_{b-i} f_{ij}$ that it contains an idle period beginning with $[k, b - i]$ and ending with $[k + j, b]$. The arrival process in A and B taken together is Poissonian at density 2λ. The idle period meant consists of $i + j$ interarrival-times of this process. Though we know that i of those interarrival-times are terminated by an arrival in B, and j by an arrival in A, this information is not relevant owing to the Markovian character of the Poisson process. The average duration of the idle period under discussion hence is $(i + j)/2\lambda$. During the $i + j$ intervals with exponential distribution that constitute this idle period, the numbers of waiting cars (both sides) are $k + b - i$, $k + b - i + 1$, $\ldots$, $k + b + j - 1$. Hence the average total waiting-time during such a period is

$$\frac{i+j}{2} \cdot (2k + 2b + j - i - 1) \cdot \frac{1}{2\lambda}.$$

The average duration of idle time T_i per gross period is

$$T_i = \sum_{k=0}^{\infty} \sum_{i=1}^{b} \sum_{j=0}^{\infty} \varphi_k q_{b-i} f_{ij} (i + j)/2\lambda,$$

or as $\sum_0^\infty \varphi_k = 1$:

$$T_i = \frac{1}{2\lambda} \sum_{i=1}^{b} \sum_{j=0}^{\infty} q_{b-i} f_{ij} (i + j). \tag{8.5.5}$$

Further the average total waiting-time $E_i(\underline{w}_{tot})$ incurred during the idle time in the gross period is (with $\sum_1^\infty k\varphi_k = \lambda$)

$$E_i(\underline{w}_{tot}) = \frac{1}{4\lambda} \sum_{k=0}^{\infty} \sum_{i=0}^{b} \sum_{j=0}^{\infty} \varphi_k q_{bi-} f_{ij} (i+j)(2k+2b+j-i-1)$$

$$= \lambda T_i + \frac{1}{4\lambda} \sum_{i=1}^{b} \sum_{j=0}^{\infty} q_{b-i} f_{ij} (i+j)(2b+j-i-1). \tag{8.5.6}$$

During a trip starting with [0, s] (probability p_s) the average number of waiting cars (both sides) linearly increases from s to $s + 2\lambda$. Hence The average total waiting-time per trip is

$$\sum_{s=0}^{\infty} (s+\lambda) p_s = \lambda + \sum_{s=0}^{\infty} s p_s.$$

The average waiting-time per car is obtained by adding the average total waiting-time per trip and that for the possible waiting-time in the idle period and then dividing this total by the average number of arrivals per gross period, i.e. by $2\lambda \cdot (1 + T_i)$:

$$w = \frac{\lambda + \sum_1^\infty s p_s + E_i(\underline{w}_{tot})}{2\lambda(1+T_i)}. \tag{8.5.7}$$

9. SIMULATION

9.1. Introduction; Non-analytical Methods

In the previous chapters the stochastic processes at hand always allowed the definition of a suitable set of states and their probabilities as well as the derivation of a set of birth-and-death equations for those quantities. Mostly, the cases chosen were so simple that the equations could be solved analytically. In more complicated situations—and real situations usually are!—though the derivation of birth-and-death equations may remain possible, those equations frequently are so awkward that they resist any analytical attack (cf. the case of Section 4.3, which must still be considered "simple").

When analytical solution of the birth-and-death equations is beyond our power, this does not necessarily mean that the method of birth-and-death equations has to be abandoned completely. There is still the possibility of numerical solution. When we confine ourselves to Erlang-k or hyperexponential distributions of holding-times, the birth-and-death equations consist of large sets of linear equations. The only difficulty is their numerousness. But one has at one's disposal the full scale of numerical methods and computer-programs based on them to cope with this difficulty. A very efficient method is to set up the birth-and-death equations for the non-stationary case and integrate those equations numerically until stationarity is reached (supposing that steady-state conditions are possible). In the case of Erlang-k and hyperexponential distributions, the set of equations for the transient behaviour is a set of linear first-order differential equations, for which very good standard numerical methods are available.

Also in more complicated cases, numerical integration of the relations describing the transient behaviour may be very well feasible. Even the seemingly awkward case of Section 4.3 can be treated in this way. When the state probabilities are dependent on the current time τ, the only change is the occurrence of extra time derivatives $\partial q_r/\partial\tau$ and $\partial q_{c+s}/\partial\tau$ in the left-hand members of (4.3.1) and (4.3.2), respectively. When at time $\tau = 0$ the system is empty, the initial conditions are all unknowns $= 0$, except $q_0(0) = 1$. Numerical integration in steps of $\Delta\tau$ does not seem to be impossible. The only real difficulty is the rapid increase of the system of equations. Hence, numerical integration will probably cease to be efficient for $c \geqq 2$ or 3, or for ϱ being too near to c.

If numerical methods are also beyond human and "computorial" power there still is a powerful method of ultimate resource, viz. *simulation*. In Section 9.2 we shall discuss the principle of *time-true simulation*, whereas Section 9.3 will describe *roulette simulation*. Subsequent sections will give some information on implementation and accuracy. Purely programmatical aspects, as well as use of special *simulation languages*, will not be dealt with. For general information about simulation cf. Tocher (1963), Naylor *et al.* (1966) and Emshoff and Sisson (1970). For special languages cf. Buxton (1968).

We shall discuss the general concept of simulation with the help of the following simple example. There are two groups of sources (A and B), producing Poisson flows of demands with rates λ_a and λ_b, respectively. There are n servers. Demands from group B are allowed to seize any free server; those demands are lost for which no free server is available on their arrival. Demands stemming from group A cannot seize a server, unless they leave at least m (>0) servers unoccupied. Once the processing of a group A demand begins, however, this action is completed. Contrary to group B demands the demands of group A have a queueing facility. One can think of the following interpretation. Let the demands of group A and B be patients to be taken into a hospital. Group B are emergency cases that may "seize" any free bed; in the case where no bed is available the patient goes to another hospital (i.e. is "lost" for the hospital under consideration). Group A are patients on a waiting list (i.e. in a queue). Their admission to hospital is made possible only in the case where m beds are still left free for emergency cases (cf. the "iconogram" of Fig. 9.1, where $n = 3, m = 1$). The model is completed by stating that the c.d.f.'s of holding-times for group A and B are $F_a(t)$ and $F_b(t)$, respectively. It is required that we determine the fraction of group B demands that is lost and the average waiting-time for group A demands. Needless to say this is a gross oversimplification of the situation at hand, but it is a good example.

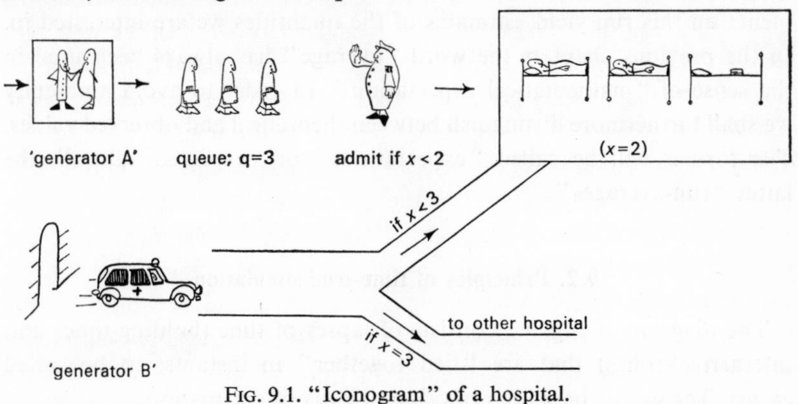

FIG. 9.1. "Iconogram" of a hospital.

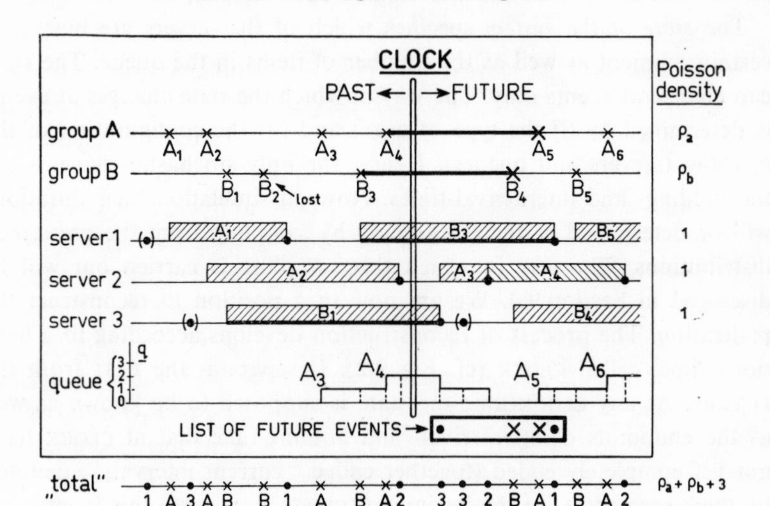

FIG. 9.2. Realization of the process of arrivals and occupations.

Let us consider the stochastic process governing the flow of demands and the occupation of servers and of the queue. One realization of such a process—i.e. one possible development of events in the system—is depicted in Fig. 9.2 for the case of Fig. 9.1 (information outside the rectangle has a bearing on Section 9.3 only). Simulation is a way of reconstructing such a realization. It is not possible to obtain more than a finite portion of such a realization, a so-called (*simulation*) *run*. Measurements on this run yield estimates of the quantities we are interested in. In the previous chapters the word "average" has always been used in the sense of "mathematical expectation". In order to avoid ambiguity we shall furthermore distinguish between theoretical and observed values. The former will be called "expectations" or "expected values", the latter "run-averages".

9.2. Principles of time-true simulation

The diagram of Fig. 9.2 consists of lapses of time (holding-times and interarrival-times) that are "tied together" in instants, to be called *events*. The events here are arrivals and ends of occupations.

The *state of the system* specifies which of the servers are busy at a certain moment as well as the number of items in the queue. The state can change at events only. The way in which the state changes at events is determined by (i) the type of event and (ii) the configuration of the *facilities* (servers and queues). Hence, the only stochastic elements are the holding- and interarrival-times. Now, in simulation their durations will be determined at their beginning by sampling from the prescribed distributions. The way in which this sampling is carried out will be discussed in Section 9.4. We are now in a position to reconstruct the realization. The process of reconstruction develops according to a fictitious time, called CLOCK (cf. Fig. 9.2). It separates the PAST from the FUTURE. At any CLOCK-time the state is supposed to be known as well as the endpoints of interarrival- and holding-time that at CLOCK have not yet completely ended (together called "current intervals", denoted by thick segments). At the beginning (CLOCK-time zero) this is satisfied, for example, by an empty system and known first arrivals in both groups.

Nothing happens until CLOCK meets the earliest of there endpoints, the *next event*. The type of this event, the state and the configuration then determine the change of state, if any, at this event. Any such a change may entail the start of one or more new current intervals, the lengths of which are immediately determined by sampling. The ends of those new current intervals mark possible new events. Hence, every event that is passed by CLOCK, may generate new *future events*. Mostly, a chronological LIST OF FUTURE EVENTS is constructed. The future events, created at the passing of an event by CLOCK, have to be inserted on this list in the right chronological place.

In early simulation practice the CLOCK was advanced in small time-steps. During most time-steps no event was passed and nothing happened. Now, the steps in which nothing happens may as well be skipped, and in current practice the CLOCK is made to jump from event to event.

During so-called simulation *runs* one can count the demands in group B as well as those that were lost. A simple division of those totals yields an estimate of the probability of loss in group B. The denominator (the total of all demands) can be replaced by its (known) expected value.

As all durations between consecutive events are also known, it is relatively easy to keep a record of the total waiting-time of group A demands during a simulation run (the area under the q-line in Fig. 9.2). Dividing this total by the total of A demands (either observed or expected) yields an estimate for the expected waiting-time of group A demands. From waiting-times of individual demands—which can be obtained with slightly more effort—a histogram of waiting-times can be constructed. This then offers an approximation to the distribution of waiting-times.

9.3. Principles of roulette simulation

The time-true simulation has three disadvantages:

(i) the implementation of the sampling from given distributions may be difficult and time-consuming;

(ii) when the system is large, much information about holding- and interarrival-times (the current intervals) should be memorized;

(iii) when the system is large, the filing of future events on their chronological list may be cumbersome or time-consuming.

In the field of telecommunication traffic research this has led to the development and widespread use of so-called *roulette simulation* (Kosten, 1942, 1948, 1970; Broadhurst and Harmston, 1949; Dietrich and Wagner, 1963; Manucci and Tonietti, 1969), that does not possess these disadvantages.

In Fig. 9.2 the arriving demands in groups *A* and *B* are given by Poisson point-processes (the crosses). Now, let the holding-times be exponentially distributed with average 1. Consequently, the arrival rates will be given in normed notation: ϱ_a and ϱ_b, respectively. Not only the interarrival-times, but also the holding-times, will now possess the property of forgetfulness (Section 1.2). A statistically true picture of holding-times may be obtained by "chopping them off" by unit density Poisson point-processes: the dots on server-occupation lines in Fig. 9.2. When such *breakdown points* occur at instances at which the associated server is not occupied (dots between brackets) nothing happens.

The five Poisson point-processes (two arrival processes and three breakdown processes) may be merged into *one* Poisson point-process, called "total". Its density is $\varrho_a + \varrho_b + 3$. In each point of this "total" process the *class* of the point (*A*, *B*, 1, 2 or 3) may be obtained by drawing lots with probabilities proportional to $\varrho_a : \varrho_b : 1 : 1 : 1$, respectively.

Now, let us suppose that we know the class indices of the consecutive points of the "total" process (marks *A*, *B*, 1, 2 or 3 on the "total" line). It is evident that one is able then to reconstruct the realization of the complete stochastic process in the system *as far as the sequence of events is concerned*! The concept "time" disappears. What remains is a so-called *sequence-true résumé* of the realization.

The class indices in question can be determined by the use of a roulette with positions marked *A*, *B*, 1, 2 and 3 in the correct proportion $\varrho_a : \varrho_b : 1 : 1 : 1$ (cf. Section 9.4 for the implementation). In a general case the roulette should possess the following sets of differently marked equiprobable positions:

(i) *breakdown positions*: one set per server; the numbers of positions in those sets are equal;

(ii) *build-up positions*: one set per group of sources; when ϱ is the arrival rate of such a group, the corresponding set of build-up positions should consist of ϱ times as many positions as the sets in (i).

When the roulette stops at a breakdown position associated with a non-occupied server, this roulette-point simply is ineffective. The idea of a roulette that is sometimes effective and sometimes not is called the *Russian roulette*. The realization of the roulette when using a computer is discussed in Section 9.4. The roulette simulation concept can also be adapted to Erlang-k distributions (Kosten, 1942 and 1948). The breakdown positions per set should now be k times as numerous. In the example the numbers of positions of the sets marked A, B, 1, 2 and 3 should then be proportional to $\varrho_a : \varrho_b : k : k : k$. A server that is occupied enters in "phase 1". Each time the roulette stops at a breakdown position associated with this server, the phase index increases by one. When ultimately this index is $k - 1$, the next occurrence of a breakdown for this server will reset the server to the "disengaged" state. Hyperexponential distributions can be coped with in a similar way.

The sequence-true résumé produced is sufficient to answer all questions that do not depend on the concept "time". The simple totalling of lost demands of group B in the example again yields an estimate for the probability of loss of group B demands.

At first sight it might appear that questions about waiting-times cannot be dealt with. This idea is not true as will be shown now. The "total" point-process is a Poisson process with some known density P, e.g. equal to the sum of the traffic densities plus k times the number of servers (in case of an Erlang-k distribution of holding-times). In the example of Fig. 9.2 we have $P = \varrho_a + \varrho_b + 3k$. The intervals between consecutive "total" points—to be called *intergeneration-times*—are exponential $(1/P)$ and are uncorrelated. Now, consider a simulation run consisting of a sequence of $N + 1$ points, marking N intergeneration-intervals. The drawings of the class indices—from which the sequence-true résumé ensues—are totally independent of the intergeneration-times. Conversely, knowledge of this résumé does not convey any information about the duration of the intergeneration-intervals. Hence, a good estimate of the total waiting-time in some queue, incurred during the simulation run, is obtained by taking the total of the N values of the queue-length assumed during the N intergeneration-times, and multiplying this total by the average intergeneration-time $1/P$. In Fig. 9.2 this means that the rectangles under the q-line, as marked by dotted lines, are taken with equal width $1/P$. The average waiting-time is obtained in the normal way, viz. by dividing the forementioned estimate of total waiting-time by the number of group A demands during the simulation run (expected or counted).

But even *distributions of waiting-times* can be obtained from the sequence-true résumé! The p.d.f. of a single intergeneration-time is $Pe^{-P\tau}$. The p.d.f. of the total duration of a string of r such intervals is obtained by convolution:

p.d.f. of r consecutive intergeneration-times

$$= (Pe^{-P\tau})^{r*} = P^r\tau^{r-1}e^{-P\tau}/(r-1)! \quad (r = 1, 2, \ldots) \tag{9.3.1}$$

This p.d.f. is *not* dependent on knowledge of the résumé.

Now, consider the traffic from some group of sources that may incur delays (say group A demands in the example). Each demand that suffers delay has to wait for some number ($\underline{r}$) of intergeneration-times (in the example: 4 of those times for both demands A_3 and A_4). Let G_r be the observed relative frequency of delayed demands that have to wait for $\underline{r} = r$ intergeneration-times [(9.3.1) gives the conditional p.d.f. for the associated waiting-time]. Then an approximation to the p.d.f. of waiting-time $\underline{W}$ per delayed demand is given by

$$\text{p.d.f.} \quad (\underline{W}) = \sum_{r=1}^{\infty} G_r P^r \tau^{r-1} e^{-P\tau}/(r-1)! \tag{9.3.2}$$

The roulette simulation ceases to be applicable when the concept "time" enters as an explicit variable. It is impossible, for example, to deal with the probability of a delay in excess of a certain prescribed duration. Furthermore, cases where the system's behaviour depends on some holding- or waiting-time exceeding some value are not covered by the roulette simulation principle. The roulette model does *not* need a list of future events, so there are *no* filing difficulties. There is *no* need for sampling from arbitrary distributions. Simple (pseudo-) random numbers are sufficient (cf. Section 9.4).

Olsson (1967) extended the use of roulette simulation to cases where arrival rates are state-dependent as in the Engset traffic model.

9.4. The implementation of randomness

In Section 9.2 we needed sampling from given distribution functions. In Section 9.3 a multi-position roulette was necessary. Both requirements are easily met when we are able to sample from uniformly distributed populations. This goal will first be considered.

In the past actual gambling devices (roulettes, dice, etc.) have been used, as well as tables of random numbers (Fisher and Yates, 1963; Kendall and Babington-Smith, 1939). These methods are very unsuitable in connection with computers. In current practice random numbers are replaced by *pseudo-random numbers.* In many instances we call numbers "random" when we are *unable or unwilling* to look into the mechanism that produces those numbers. In this respect it is immaterial whether those numbers are produced by a physical randomizing device or in a mathematical way. For example, Fisher and Yates' tables consisted of 15th, ..., 19th digits of logarithms. Though there is nothing random in the actual sequence of logarithms, the mere fact that we ignore the part of the table from which those five decimal numbers stem makes us accept them as "random numbers". Computers usually handle numbers of some fixed length, say of m digits (e.g. $m = 9$ decimals or $m = 32$ binary digits). Those standard length numbers are called "words" (in view of the fact that they also may represent non-numerical information). Computers mostly have one or more subprogrammes called *pseudo-random number generators.* Each time such a pseudo-random number generator is called for, it produces one word of standard length, to be used as a substitute for a random number.

Many of these pseudo-random number generators incorporate a "black box" carrying out an operation of the type:

$$x_n = f(x_{n-1}). \qquad (9.4.1)$$

This means that any new result x_n of the generator is produced by performing some given operation $f(...)$ on the former result x_{n-1}. The generator is *set* initially by procuring a so-called *seed* x_0. In other pseudo-random number generators new pseudo-random numbers are constructed from *two* predecessors,

$$x_n = f(x_{n-1}, x_{n-2}), \tag{9.4.2}$$

in which case two seeds are necessary.

Usually pseudo-random numbers are taken to be non-negative. Single length words consist of a finite number (m) of digits (decimal or binary). Hence, there are a finite number of different words ($M = 10^m$ or 2^m, respectively). This means that a generator of type (9.4.1) in at most M operations will produce a word that has already been encountered. Then *cycling* occurs, i.e. exact repetition of the history. The length of the cycle p, called the *period*, is $\leqq M$. For generators of type (9.4.2) the period is $\leqq M^2$, as there are M^2 different word-pairs. It will be clear that too short a period is inacceptable.

The main conditions to be satisfied by a pseudo-random number generator are:

(i) a large period,
(ii) uniformity of distribution,
(iii) freedom from correlation of successive outcomes.

If the generator has maximum period [M for (9.4.1), M^2 for (9.4.2)], each number $0, \ldots, M-1$ appears with equal frequency and the uniformity of the distribution is assured. If not, the uniformity must be tested. One can split up the range $(0, \ldots, M-1)$ in, say, 100 intervals of (approximately) equal length and test for uniformity, e.g. by the χ^2-test, on those 100 classes. A test on correlation should always be carried out.

The pseudo-random number generators that have found widespread use are:

(i) *The mixed congruence generator*:

$$x_n = ax_{n-1} + c \pmod{M} \quad (a, c > 0). \tag{9.4.3}$$

The former number x_{n-1} is multiplied by a fixed single length number a, a constant c is added and the "tail" of the double-length result is taken as x_n. The generator described has the great advantage that under favourable conditions the period is maximal (M), thus ensuring uniformity. Those conditions are (cf. Hammersley and Handscomb, 1967, p. 28):

c and M are mutually prime,
$a \equiv 1$ (mod. p) for every prime factor p of M,
$a \equiv 1$ (mod. 4) if 4 divides M.

(ii) *The Fibonacci generator*:

$$x_n = x_{n-1} + x_{n-2} (\text{mod. } M). \tag{9.4.4}$$

This generator is very fast, as the operation of addition may be very much quicker on a computer than multiplication. Under favourable conditions (e.g. $x_0 = x_1 = 1$) the period is $\frac{3}{2}M$. Uniformity is met reasonably well.

Freedom from correlation is a condition that is not always met so very well (especially the Fibonacci generator is known to possess some correlation). The following measure (attributed to Page) has been shown to remove any correlation drastically. A buffer with (say) 128 single word memory locations is inserted between the generator to be improved and the outer world. Initially those 128 positions are filled by generator-words. From then on pairs of generator-words (say x_{2n} and x_{2n+1}) are used in combination to form pseudo-random numbers by the following two-stroke operation:

(i) a word from the buffer is forwarded to the outer world as a pseudo-random number; the buffer location i ($i = 0, \ldots, 127$) from which to take this word is designated by a simple operation (in a binary computer: a shift of seven positions):

$$i = [128 x_{2n}/M]; \tag{9.4.5}$$

(ii) this ith location is refilled with x_{2n+1}.

Assuming that computing the generator-words takes most of the time, we can say that correlation has been practically removed at the expense of some memory space and a reduction in speed by a factor 2. For pseudo-random number generators in general cf. Jansson (1966).

From now on we shall consider the words issued by the generator (either directly or via a buffer) as representing m digits of a true fraction (i.e. pseudo-random numbers are "reduced" by division by M). Hence, those (reduced) pseudo-random numbers, furthermore, are $\geqq 0$ and <1.

A roulette is implemented very easily. Let the "total" point-process consist of points of S classes with probabilities $p_1, \ldots, p_s$ ($\sum p_i = 1$). Then the interval [0, 1) is divided into the following S subintervals I_i:

$$I_i = \left[\sum_1^{i-1} p_j, \sum_1^{i} p_j\right) \quad (i = 1, \ldots, S). \tag{9.4.6}$$

Whenever a reduced pseudo-random number is in the interval I_i, it marks a point of class i.

The pseudo-random numbers may also be used to implement sampling from some given population with c.d.f. $F(t)$. Let u be some reduced pseudo-random number. Then the sample value t is determined in such a way that $u = F(t)$. For when u and t are related in this way, the fraction of pseudo-random numbers not exceeding u and the fraction of sample values not exceeding t are equal (viz. u for any value of u). Determining t from $u = F(t)$ is a process of *inverse interpolation*. This is a rather time-consuming numerical process. It can best be implemented by a piecewise linearization of the c.d.f. The computer then needs a set of coordinate-pairs for the ends of the linear pieces and a subprogramme for carrying out the interpolation on the pieces.

In the opinion of the present author the following scheme is very adequate. The continuous distribution in question is substituted for by a discrete distribution with equal probabilities. For this purpose the interval $t = [0, \infty)$ is, for instance, divided in the ten *decile classes*, all of which have the class-probability 1/10 (cf. Fig. 9.3). The averages of the ten decile-classes $d_0, \ldots, d_9$ are determined. Now, a reduced pseudo-random number u is drawn. Then an index i is determined from

$$i = [10u]. \tag{9.4.7}$$

This index i takes on values 0, ..., 9 with equal probabilities. The sample value for the discretized distribution then is put on $t = d_i$.

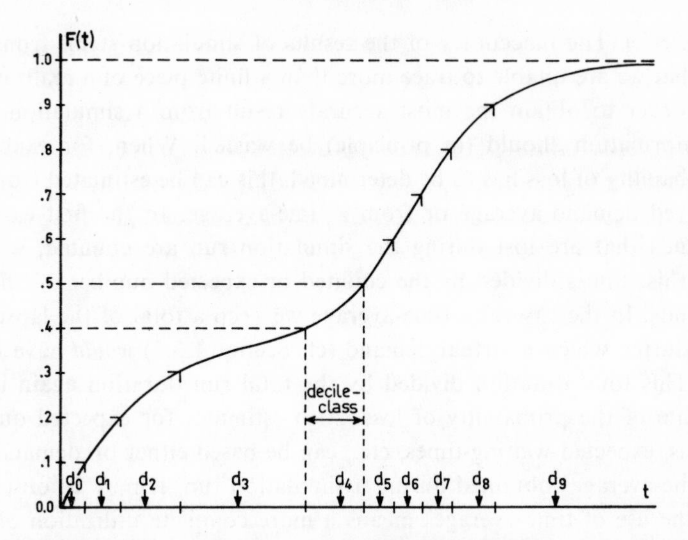

FIG. 9.3. Substitution for a continuous distribution by the uniform discrete distribution of decile class average.

It will be obvious that the process described may be very much quicker than inverse interpolation on the (linearized) c.d.f. Moreover, the discretized distribution has the same average as the original distribution, which is *not* the case with the piecewise linearized function unless special measures have been taken.

If a discretization in ten decile classes is thought to be too crude, 100 (per)centile classes can be introduced at the mere expense of some memory space.

9.5. Accuracy of simulation results

Simulation is a way of measuring in a hypothetical system. Just as in the case of actual measurements, results are liable to errors. Those errors may be decreased by repetition of measurements.

The quantities we want to determine by simulation mostly are probabilities (of loss, of delay, etc.) or mathematical expectations (of waiting-times, etc.). The inaccuracy of the results of simulation stems from the fact that we are unable to trace more than a finite piece of a realization.

In order to obtain the most accurate result from a simulation run, no information should (in principle) be wasted. When, for example, a probability of loss has to be determined, this can be estimated from an observed demand-average or from a time-average. In the first case all demands that are lost during the simulation run are counted, whereafter this sum is divided by the counted or expected number of offered demands. In the case of a time-average we keep a total of the lapses of time during which a virtual demand (cf. Section 1.3.2) *would* have been lost. This total duration divided by the total run duration again is an estimate of the probability of loss. Also estimates for expected queue-lengths, expected waiting-times, etc., can be based either on demand- or on time-averages obtained during a simulation run. It may be observed that the use of time-averages means a more complete utilization of the information contained in the realization than the use of demand-averages. Hence, the estimates obtained from time-averages will be better. Confirmation of this conjecture has been found analytically for the standard cases ($M/M/c$-blocking: cf. Kosten *et al.*, 1949; $M/M/c$-delay: cf. Kosten, 1952).

It is good practice to divide a long simulation run (say of 10,000 demands) into a number of subruns (say 10 times 1000 demands). Both the long run and the series of subruns will of course yield the same observed average. When using results of subruns, however, the variance for subrun results can be estimated as well. *Assuming the results of successive subruns not to be correlated*, the variance of the average subrun result would be obtained by dividing the variance for the subruns by the number of subruns. This then would lead to *confidence limits* for the final result. Now, there is one pitfall to be avoided. As the system's condition at the end of one subrun at the same time is the initial condition for the next subrun, consecutive subrun results are mutually dependent (with positive correlation). This dependence will be the heavier the shorter the subruns: in short subruns the influence of the initial condition has not yet faded away at the end. When this correlation is neglected, the confidence limits obtained may be far too optimistic! This should constitute a warning against subdividing the total run too much.

Mechanic and McKay (unpublished IBM report) have devised a technique for automatic adjustment of subrun size.

The correlation between results of consecutive subruns also comes to light in the *clustering effect* of lost or delayed demands and of periods with large values of queuelength, etc. This point will be illustrated in Section 9.6 by an example.

A realization necessarily starts with some arbitrary state of the system (e.g. an "empty system"). When the aim of the simulation is to estimate steady-state expected values, it is a good policy to disregard the results of the first subrun(s), in order to bring the system into a state that is not too peculiar with a view to stationarity.

Simulation is a rather expensive technique. Moreover, accuracy increases with the square root of total run length only. Hence, the cost increases rather heavily with the degree of accuracy wanted. For this reason it has been proposed to use *variance reducing techniques* (Hammersley and Handscomb, 1967; Hillier and Lieberman, 1967; Kleynen, 1971). Their use is invariably demonstrated with the help of some very simple example, e.g. a single server situation.

The present author does not feel confident that these techniques are so very useful in complicated cases, though he must confess that this opinion is not based on experiments. When using time-true simulation the technique of *antithetic variables* could be tried in any case. Its philosophy is the following. When u is some pseudo-random number between 0 and 1 (i.e. a sample from a uniform distribution on $[0, 1)$), the quantity $1 - u$ also is such a pseudo-random number. When u is used to construct a sample from some c.d.f. $F(t)$—say for determining a holding-time or an interarrival-time—large (small) values of u will result in large (small) values of t. Now, it is possible to have some run repeated in exactly the same way as originally, but using the complements $1 - u$ of all pseudo-random numbers u (antithetic variables). Every time a sampled holding-time (or interarrival-time) in the first run is relatively large, the corresponding time in the second run is small, and conversely. This means, for example, that when an estimate for an average waiting-time from the results in the first run is too large (supposedly by the holding-times being relatively large and/or the interarrival-times relatively short), the estimate based on the second run probably will be too small, and conversely. It may be conjectured that the average of both estimates is better than the averages of two estimates from unconnected runs. Whether the technique pays or not, it may be used without extra memory space or computer time. Hence, its use is to be advocated.

Often simulation is used on a family of system models in separate series of experiments, e.g. with varying number of servers, etc. As the use of pseudo-random number generators renders the simulation deterministic, it is possible to employ exactly the same arrival patterns and holding-times for the different experiments. This means that large and small outcomes in the different experiments will be correlated. The difference between results of two series of experiments will be more accurate than the outcomes of the series themselves. The so-called *residual variance* is smaller than the variance of the results in each series.

The large cost of simulation cries out for appropriated *designed experiments* (cf. Cochran and Cox, 1957). As the experiments are much better controllable than even the best physical experiments in the laboratory, design of experiment is very well feasible. Much work needs be done in this area.

9.6. The "cluster effect"

As has already been emphasized in Section 9.5, lost and delayed demands tend to occur in clusters, as do periods with large queue-length values, etc. This implies a positive correlation between results of consecutive simulation-runs. In the following simple example the influence of this correlation on the accuracy of measurements has been traced analytically.

Consider the system $M/M/1$-delay with arrival rate ϱ. Suppose stationarity to hold. Denote by $\underline{r}$ the number of demands in the system encountered by an arriving demand (inclusive of the one being served, if any; not counting the arriving demand). Let $\underline{s}(\tau)$ be the sum of the $\underline{r}$-values that are encountered by arrivals during $(0, \tau)$. Then $\underline{s}(\tau)/\varrho\tau$ is an estimate for the expected number $E(\underline{r})$ of items in the system. It is the aim of this section to calculate var $\{\underline{s}(\tau)\}$ and to estimate the accuracy of the (fictitious) measurement of $E(\underline{r})$, described above.

As the system is supposed to be in steady-state condition at $\tau = 0$ the probability p_r of $\underline{r}$ having the value r and its generating function are:

$$p_r = (1 - \varrho)\,\varrho^r \triangleq \frac{1 - \varrho}{1 - \varrho x}. \tag{9.6.1}$$

Hence,

$$\left.\begin{aligned} E(\underline{r}) &= \sum_0^\infty r p_r = x \frac{d}{dx} \frac{1 - \varrho}{1 - \varrho x}\bigg|_{x=1} = \frac{\varrho}{1 - \varrho}, \\ E(\underline{r}^2) &= x \frac{d}{dx}\, x \frac{d}{dx} \frac{1 - \varrho}{1 - \varrho x}\bigg|_{x=1} = \frac{\varrho(1 + \varrho)}{(1 - \varrho)^2}, \\ \text{var}\,(\underline{r}) &= E(\underline{r}^2) - E^2(\underline{r}) = \frac{\varrho}{(1 - \varrho)^2}. \end{aligned}\right\} \tag{9.6.2}$$

As the expected number of arrivals in $(0, \tau)$ is $\varrho\tau$ one has

$$E\{\underline{s}(\tau)\} = \varrho\tau E(\underline{r}) = \frac{\varrho^2\tau}{1 - \varrho}. \tag{9.6.3}$$

For small values of τ it is highly improbable that more than one demand arrives during $(0, \tau)$, whilst the probability of just one arrival is $\approx \varrho\tau$. Hence,

$$\text{var}\,\{\underline{s}(\tau)\} \approx \varrho\tau \cdot E(\underline{r}^2) - E^2\{\underline{s}(\tau)\} \approx \frac{\varrho^2(1 + \varrho)\,\tau}{(1 - \varrho)^2} \quad (\tau \ll 1). \tag{9.6.4}$$

If the observed $\underline{s}$ values of adjacent intervals were uncorrelated, both $E\{\underline{s}(\tau)\}$ and var $\{\underline{s}(\tau)\}$ would be linearly dependent on τ, and

$$Q(\tau) := \frac{\text{var}\,\{\underline{s}(\tau)\}}{E\{\underline{s}(\tau)\}} \tag{9.6.5}$$

would be independent of τ. Its (constant) value then would result from (9.6.3 and 9.6.4) by taking $\tau \to 0$:

$$Q(0) = \frac{1 + \varrho}{1 - \varrho}. \tag{9.6.6}$$

We shall presently calculate the actual value of $Q(\tau)$ by analytical and numerical means. It turns out to be far from constant.

Let $[r, s]$ be the state of the system with $\underline{r} = r$ demands (including the one being served, if any), whilst the sum $\underline{s}$, defined above, amounts to s. The probability of the system being in $[r, s]$ at time τ will be denoted by $u_{rs}(\tau)$. When $\underline{r}$ changes from $r - 1$ to r by a new arrival, $\underline{s}$ increases by $r - 1$. Hence, the scheme of transitions is:

State at time τ	Probability of transition	State at time $\tau + \Delta\tau$
$[r - 1, s - r + 1]$	$\varrho\Delta\tau$ $\longrightarrow$	$[r, s]$
$[r, s]$	$1-\varrho\Delta\tau-\Delta\tau$ $\longrightarrow$ (last term to be deleted for $r = 0$)	$[r, s]$
$[r+ 1, s]$	$\Delta\tau$ $\longrightarrow$	$[r, s]$

From this scheme we obtain the following set of generalized birth-and-death equations:

$$\frac{d}{d\tau} u_{rs} = \varrho u_{r-1,s-r+1} - (\varrho + 1) u_{rs} + u_{r+1,s} + \delta_r^0 u_{0s} \qquad (u_{-1,s} \equiv 0;\ r, s = 0, 1, 2, \dots). \tag{9.6.7}$$

In view of the stationarity the initial conditions are

$$u_{rs}(0) = p_r \delta_s^0 \quad (r, s = 0, 1, 2, \dots). \tag{9.6.8}$$

Let

$$\mu_r(\tau) := \sum_{s=0}^{\infty} s u_{rs}(\tau); \quad \nu_r(\tau) := \sum_{s=0}^{\infty} s^2 u_{rs}(\tau). \tag{9.6.9}$$

On account of the stationarity we have

$$\sum_{s=0}^{\infty} u_{rs}(\tau) = p_r = (1 - \varrho) \varrho^r. \tag{9.6.10}$$

Multiplication of (9.6.7) by s, summing over s and application of (9.6.9 and 9.6.10) yields

$$\frac{d}{d\tau}\mu_r = \varrho\mu_{r-1} - (\varrho + 1)\mu_r + \mu_{r+1} + \delta_r^0\mu_0 + \varrho(r-1)p_{r-1}$$
$$(r = 0, 1, 2, \ldots). \qquad (9.6.11)$$

Multiplication by s^2 and summation analogously yields

$$\frac{d}{d\tau}\nu_r = \varrho\nu_{r-1} - (\varrho + 1)\nu_r + \nu_{r+1} + \delta_r^0\nu_0 + 2\varrho(r-1)\mu_{r-1}$$
$$+ \varrho(r-1)^2 p_{r-1} \quad (r = 0, 1, 2, \ldots). \qquad (9.6.12)$$

This system of simultaneous differential equations may be integrated, with the initial conditions:

$$\mu_r(0) = \nu_r(0) = 0 \quad (r = 0, 1, 2, \ldots). \qquad (9.6.13)$$

One can use, for example, numerical integration by the Runge–Kutta method. The systems are truncated at some large index value $r = R$. The quantities μ_{R+1} and ν_{R+1}, occurring in the differential equations for $r = R$, should be taken to be identically zero.

After a numerical integration has been performed, we can evaluate for any τ:

$$E\{\underline{s}(\tau)\} = \sum_0^\infty \mu_r(\tau), \qquad (9.6.14\text{a})$$

$$\text{var}\{\underline{s}(\tau)\} = \sum_0^\infty \nu_r(\tau) - E^2\{\underline{s}(\tau)\}, \qquad (9.6.14\text{b})$$

and the related $Q(\tau)$ can be determined. Summation of (9.6.11) for $r = 0, 1, 2, \ldots$ yields

$$\frac{d}{d\tau}E\{\underline{s}(\tau)\} = \varrho\sum_0^\infty (r-1)p_{r-1} = \frac{\varrho^2}{1-\varrho}$$

which result is concordant with (9.6.3). The numerical result (9.6.14a) may be used as a check.

The described procedure has been applied to the case $\varrho = 0.5$. It appears that $Q(\tau)$ indeed increases rather strongly when τ increases, indicating positive correlation. For $\tau \to \infty$ the limiting value $Q(\infty)$ may be obtained analytically.

When generating functions and Laplace Transforms are introduced (cf. Appendix),

$$\begin{aligned} \mu_r(\tau) &\triangleq M(x, \tau) \doteqdot \mathfrak{M}(x, z), \\ \nu_r(\tau) &\triangleq N(x, \tau) \doteqdot \mathfrak{N}(x, z), \end{aligned} \tag{9.6.15}$$

(9.6.11), (9.6.12) and (9.6.13) yield

$$D(x, z)\,\mathfrak{M}(x, z) = (1 - x)\,\mathfrak{M}(0, z) - \frac{\varrho^2(1 - \varrho)}{z} \frac{x^3}{(1 - \varrho x)^2}, \tag{9.6.16}$$

$$D(x, z)\,\mathfrak{N}(x, z) = (1 - x)\,\mathfrak{N}(0, z) - 2\varrho x^3 \frac{\partial}{\partial x}\mathfrak{M}(x, z) - \frac{(1 - \varrho)\,\varrho^2}{z} x^3 \frac{1 + \varrho x}{(1 - \varrho x)^3}, \tag{9.6.17}$$

with

$$D(x, z) := \varrho x^2 - (\varrho + z + 1)\,x + 1. \tag{9.6.18}$$

When $D(x, z)$ is considered as a function of x, one of the zeros—say $x = \xi(z)$—approximates 1 for $z \to 0$:

$$\xi = \xi(z) = 1 - \frac{z}{1 - \varrho} + \frac{z^2}{(1 - \varrho)^3} - \frac{1 + \varrho}{(1 - \varrho)^5} z^3 + \cdots \tag{9.6.19}$$

As $|\xi| < 1$ for $\operatorname{Re} z = \delta > 0$, and as $\mathfrak{M}(x, z)$ should remain finite for $|x| < 1$, $\operatorname{Re} z > 0$, the right-hand member of (9.6.16) must vanish for $x = \xi$, and hence,

$$\mathfrak{M}(0, z) = \frac{\varrho^2(1 - \varrho)\,\xi^3}{z(1 - \varrho\xi)^2\,(1 - \xi)} = \frac{\varrho(1 - \varrho)}{z^3}\{1 - (1 + z)\,\xi\}. \tag{9.6.20}$$

With this expression $\mathfrak{M}(x, z)$ is known from (9.6.16) and then $\mathfrak{N}(x, z)$ from (9.6.17).

The expressions are rather unwieldy. Nevertheless, it is possible to obtain a series expansion for $\mathfrak{N}(1, z)$ with the aid of (9.6.19):

$$\mathfrak{N}(1, z) = \frac{2\varrho^4}{(1 - \varrho)^2 z^3} + \frac{\varrho^2(1 + 5\varrho - 3\varrho^2 + \varrho^3)}{(1 - \varrho)^4 z^2} + O(z^{-1}). \tag{9.6.21}$$

Hence,
$$\text{var}\{\underset{\sim}{s}(\tau)\} = N(1,\tau) - E^2\{\underset{\sim}{s}(\tau)\} \doteq \mathfrak{N}(1,z) - \frac{2\varrho^4}{(1-\varrho)^2 z^3}$$
$$= \frac{\varrho^2(1+5\varrho-3\varrho^2+\varrho^3)}{(1-\varrho)^4 z^2} + O(z^{-1}).$$

Hence we obtain the asymptotic approximation for $\tau \to \infty$:

$$\text{var}\{\underset{\sim}{s}(\tau)\} \sim \frac{\varrho^2(1+5\varrho-3\varrho^2+\varrho^3)\,\tau}{(1-\varrho)^4}, \tag{9.6.22}$$

and

$$Q(\infty) = \frac{1+5\varrho-3\varrho^2+\varrho^3}{(1-\varrho)^3}. \tag{9.6.23}$$

For $\varrho = 0.5$ the results are:

τ	$Q(\tau)$
0	3
10	12.25
20	16.12
30	18.10
40	19.24
50	19.97
...	...
...	...
∞	23

The values $Q(0) = 3$ and $Q(\infty) = 23$ stem from (9.6.6) and (9.6.23), respectively. When the subruns are taken with length $\tau = 10$ we have $Q(10) = 12.25$, i.e. about half the value for $\tau \to \infty$. This means that taking subruns of length 10 and estimating the accuracy from the variance of subrun result would lead to confidence limits that are too optimistic by a factor 1.4. For $\varrho \to 1$ the influence of this effect becomes greater (for $\varrho = 0.9$ one has $Q(0) = 19$; $Q(\infty) = 3799$).

The derivation given above is to a certain extent a computer-aided parallel to earlier analytical work by Kosten *et al.* (1949) and Kosten (1952).

APPENDIX

GENERATING FUNCTIONS AND LAPLACE TRANSFORMS

AN *Arithmetic Function* (AF) is a function that is defined for non-negative integer values of its argument only. It will be denoted by f_r, g_s, φ_r, etc., where r (or s) takes on the values 0, 1, 2, ... When the value of an AF is not formally defined (as, for example, f_{r-1} for $r = 0$) it is assumed to have the value 0. On the other hand, an AF is never defined for other values of arguments than 0, 1, 2, ... So the AF f_{r+1} is *not* defined for $r = -1$, though f_r may have a well-defined value for $r = 0$.

An AF f_r will be said to be "of limited growth α" if there exist positive α and M such that $|f_r| < M\alpha^r$ ($r = 0, 1, 2, \ldots$). If f_r is of limited growth α, the series $\sum_{r=0}^{\infty} f_r x^r$ possesses a radius of convergence $\geqq 1/\alpha > 0$. Within the domain of convergence it determines an analytic function $F(x) := \sum_{r=0}^{\infty} f_r x^r$ which will be called the *generating function* (GF) of f_r.

Conversely, when $F(x)$ is analytic in $x = 0$, it may be expanded into a power series in x and the coefficients $f := F^{(r)}(0)/r!$ form an AF that is of limited growth.

The one-to-one correspondence between AF and GF will be denoted by the symbol $\triangleq$. As far as possible corresponding AF's and GF's will be denoted by corresponding lower and upper case characters, respectively. Only in the case where the AF is a cumulative distribution function, will both AF and GF be denoted by the same upper-case character. Moreover, if the AF has an argument r (or s), the GF will be given x

(or y) as a corresponding argument. So we shall write

$$f_r \quad \triangleq F(x) \text{ [i.e. } F(x) = \sum f_r x^r ; f_r = \frac{1}{r!} F^{(r)}(0)],$$

$$g_s \quad \triangleq G(y),$$

$$\varphi_r \quad \triangleq \Phi(x),$$

$$p_r(\tau) \triangleq P(x, \tau) \quad (\tau \text{ parameter})$$

and occasionally

$$f_{rs} \triangleq \triangleq F(x, y) \quad \text{to denote} \quad F(x, y) = \sum_{r=0}^{\infty} \sum_{s=0}^{\infty} f_{rs} x^r y^s.$$

When the f_r are probabilities of a set of mutually exclusive events, the GF $F(x)$ exists.

When $f_r \triangleq F(x)$, $f_r(\tau) \triangleq F(x, \tau)$ and $g_r \triangleq G(x)$, the following rules can be verified:

Rule		Comment	
af_r	$\triangleq aF(x)$	linearity of transformation	(A1)
$f_r + g_r$	$\triangleq F(x) + G(x)$		(A2)
f_{r+1}	$\triangleq \{F(x) - F(0)\}/x$		(A3)
f_{r-1}	$\triangleq xF(x)$		(A4)
$\sum_{i=0}^{r} f_i g_{r-i}$	$\triangleq F(x)\, G(x)$	("convolution sum")	(A5)
rf_r	$\triangleq x \dfrac{dF}{dx}$		(A6)
f_r^1	$\triangleq F(x)/(1 - x)$	$f_r^1 := \sum_{i=0}^{r} f_i$ ("first sum")†	(A7)
f_r^n	$\triangleq F(x)/(1 - x)^n$	$f_r^n := \sum_{i=0}^{r} f_i^{n-1}$ ("nth sum")†	(A7′)
$\dfrac{df_r(\tau)}{d\tau}$	$\triangleq \dfrac{\partial}{\partial \tau} F(x, \tau)$	(Under certain conditions; cf. Smirnov, 1964)	(A8)

† The upper index does not signify a power. Powers of AF's to be denoted as follows: $\varphi_r^1 \cdot \varphi_r^1 = (\varphi_r^1)^2$, etc.

Provided that the sums converge the following relations hold:

$$\sum_{r=0}^{\infty} f_r = F(1), \tag{A 9}$$

$$\sum_{r=m}^{\infty} r(r-1)\dots(r-m+1) f_r = F^{(m)}(1). \tag{A 9'}$$

In case the f_r are the probabilities of a complete set of events the left-hand member of (A 9′) is termed the *m-th factorial moment* of f_r. Its importance stems from the fact that it may be easily computed in the way shown when the GF $F(x)$ is known. The ordinary moments can be obtained from those factorial moments:

$$E(\underline{r}) = \sum r f_r = F'(1), \tag{A 10}$$

$$E(\underline{r}^2) = \sum r(r-1) f_r + \sum r f_r = F''(1) + F'(1), \tag{A 10'}$$

etc.

In order to use the given rules, some standard cases of GF's must be known. The following results may be easily verified:

$$\delta_r^0 \triangleq 1 \quad \text{(Kronecker symbol; } \delta_r^k = 1 \text{ if } r = k, \text{ otherwise } 0\text{)}, \tag{A 11}$$

$$(\pm 1)^r \frac{n(n-1)\dots(n-r+1)}{r!} \triangleq (1 \pm x)^n, \tag{A 12}$$

$$\varphi_r := \frac{e^{-\varrho r}}{r!} \triangleq e^{-\varrho(1-x)} \quad \text{(Poisson distribution).} \tag{A 13}$$

The following is a simple example of the use of the rules. It is required to solve the set of equations

$$\varrho p_{r-1} - (\varrho + r) p_r + (r+1) p_{r+1} = 0 \quad (r = 0, 1, \dots, c-1) \tag{2.1.2}$$

under the condition

$$\sum_{0}^{c} p_r = 1. \tag{2.1.5}$$

The treatment in Section 2.1 uses the (very simple) trick of summing the equations (cf. 2.1.4). By the use of GF's the solution may be given without any trick. Suppose (2.1.2) to be valid for $r = 0, 1, \dots, \infty$. The additional equations for $r = c, c + 1, \ldots$ define a number of quantities $p_{c+1}, p_{c+2}, \ldots$ which have mathematical meaning only. Now, p_r is an AF. Let $p_r \triangleq P(x)$. Combining (A 6) and (A 3) yields

$$(r + 1)\, p_{r+1} = (rp_r)_{r+1 \to r} \triangleq \frac{1}{x}\left[x\frac{dP}{dx} - \left(x\frac{dP}{dx} \right)_{x=0} \right] = \frac{dP}{dx}. \qquad \text{(A14)}$$

Together with (A 1, 2 and 6) this then yields the GF of the left-hand member of (2.1.2), which should be zero:

$$\varrho xP - \varrho P - x\frac{dP}{dx} + \frac{dP}{dx} = 0$$

or

$$\frac{dP}{dx} = \varrho P.$$

Its solution is $P(x) = Ke^{-\varrho(1-x)}$, which generates (cf. A 13)

$$p_r = K\varphi_r. \qquad \text{(A 15)}$$

Now (cf. A 7), $p_c^1 = \sum_0^c p_r = K\varphi_c^1$. Insertion in (2.1.3) yields K, and hence,

$$p_r = \varphi_r / \varphi_c^1 \qquad \text{(A 16)}$$

which is the solution sought.

In the identity

$$\frac{F(x)}{(1-x)^n} = \frac{xF(x)}{(1-x)^n} + \frac{F(x)}{(1-x)^{n-1}}$$

the three constituents are GF's of f_r^n, f_{r-1}^n and f_r^{n-1}, respectively. Hence,

$$f_r^n = f_{r-1}^n + f_r^{n-1}. \qquad \text{(A 17)}$$

Now, let us consider the nth sums φ_r^n of φ_r. Apart from (A 17)

$$\varphi_r^n = \varphi_{r-1}^n + \varphi_r^{n-1} \qquad \text{(A 17')}$$

another recursion-formula may be obtained by translating the identity:

$$\frac{d}{dx}\frac{e^{-\varrho(1-x)}}{(1-x)^n} = \varrho\frac{e^{-\varrho(1-x)}}{(1-x)^n} + \frac{n^{-\varrho(1-x)}}{(1-x)^{n+1}}.$$

This yields (cf. A 14)

$$(r+1)\varphi_{r+1}^{n} = \varrho\varphi_{r}^{n} + n\varphi_{r}^{n+1}. \tag{A 18}$$

As an application let us transform the expression (2.3.6)

$$D = \frac{\varphi_c c/(c-\varrho)}{\varphi_{c-1}^{1} + \varphi_c \cdot c/(c-\varrho)} = \frac{c\varphi_c}{c(\varphi_{c-1}^{1} + \varphi_c) - \varrho\varphi_{c-1}^{1}}$$

$$= \frac{c\varphi_c}{c\varphi_c^{1} - \varrho\varphi_{c-1}^{1}} = \frac{c\varphi_c}{\varphi_{c-1}^{2}}. \tag{A 19}$$

When (A 5) is used with $G = (1-x)^{-n}$, and combined with (A 7′) and (A 12) we obtain

$$f_r^n = \sum_{i=0}^{r} \frac{n(n+1)\dots(n+i-1)}{i!} f_{r-i}, \tag{A 20}$$

which expresses the nth sum in terms of the unsummed AF. Now, we shall furthermore use (A 20) as a defining expression for "nth sums" in cases where n takes on other values than 0, 1, 2, ... (possibly complex). When $n = 0, 1, 2, \dots$, the definition is consistent with the original sum definition. Those generalized sums then obey all the results given for the simple sum-function, given above. In particular, (A 17′) and (A 18) remain valid for any complex value of n.

As another example consider the derivation of (2.3.8). The sum $\sum_{s=0}^{\infty} s(\varrho/c)^s$ may be obtained as follows:

$$(\varrho/c)^s \triangleq (1-\varrho x/c)^{-1},$$

$$s(\varrho/c)^s \triangleq x\frac{d}{dx}(1-\varrho x/c)^{-1} = \varrho cx(c-\varrho x)^{-2},$$

$$\sum_{s=0}^{\infty} s(\varrho/c)^s = |\varrho cx(c-\varrho x)^{-2}|_{x=1} = c\varrho/(c-\varrho)^2. \tag{A 21}$$

The use of *Laplace Transforms* (LT) to represent real-valued *original functions* (OF) defined on $[0, \infty)$ is so well known that we need not say much about it (cf., for example, Carslaw and Jaeger, 1953).

The LT will be given the function symbol of the OF, printed in gothic character. The argument of the OF will normally be τ (or occasionally u), whilst in the LT z will be used. The symbol of correspondence will be $\fallingdotseq$

So we shall write

$$f(\tau) \risingdotseq \mathfrak{f}(z); \quad p_r(\tau) \risingdotseq \mathfrak{p}_r(z); \quad P(x, \tau) \risingdotseq \mathfrak{P}(x, z) \tag{A 22}$$

to denote

$$\mathfrak{f}(z) = \int_0^\infty e^{-z\tau} f(\tau)\, d\tau \quad \text{(etc.)} \tag{A 23}$$

and conversely

$$f(\tau) = \frac{1}{2\pi i} \int_{\lambda - i\infty}^{\lambda + i\infty} e^{z\tau} \mathfrak{f}(z)\, dz \quad \text{(etc.)}. \tag{A 24}$$

When the OF $f(\tau)$ does not increase more rapidly than exponentially, i.e. $|f(\tau)| < Me^{\alpha\tau}$, the LF (A 23) exists for $\operatorname{Re} z > \alpha$. If, moreover, $f(\tau)$ possesses a continuous derivative, the inversion integral (A 24) exists for $\lambda > \alpha$ and it determines the OF $f(\tau)$.

When $f(\tau)$ is the p.d.f. of the stochastic variable $\underline{\tau}$, the LT is equal to $\mathfrak{f}(z) = E(e^{-z\underline{\tau}})$; it converges for $\operatorname{Re} z \geqq 0$. The moments of the distribution are connected with the derivatives of the LT in $z = 0$:

$$E(\underline{\tau}^m) = (-1)^m\, \mathfrak{f}^{(m)}(0). \tag{A 25}$$

Let $f_1(\tau)$ and $f_2(\tau)$ be the p.d.f.'s of two uncorrelated stochastic variables $\underline{\tau}_1$ and $\underline{\tau}_2$. Let $f(\tau)$ be the p.d.f. of the sum $\underline{\tau} = \underline{\tau}_1 + \underline{\tau}_2$.

Introduce the LT's:

$$f_1(\tau) \risingdotseq \mathfrak{f}_1(z); \quad f_2(\tau) \risingdotseq \mathfrak{f}_2(z); \quad f(\tau) \risingdotseq \mathfrak{f}(z).$$

Then we have:

$$\mathfrak{f}z = E(e^{-\underline{\tau}_1 - \underline{\tau}_2}) = E(e^{-\underline{\tau}_1})\, E(e^{-\underline{\tau}_2}) = \mathfrak{f}_1(z)\, \mathfrak{f}_2(z) \tag{A 26}$$

Hence, the LT of the p.d.f. of the sum of two uncorrelated stochastic variables equals the product of the LT's of the p.d.f.'s. of the components. The p.d.f. of the sum itself is equal to the so-called *convolution-integral*, sometimes called the "symbolic or asterisk product":

$$f(\tau) = \int_0^\tau f_1(u) f_2(\tau - u)\, du = \int_0^\tau f_1(\tau - u) f_2(u)\, du := f_1(\tau) * f_2(\tau). \tag{A 27}$$

The p.d.f. of the sum of n uncorrelated components with p.d.f.'s. $f_1(\tau), \ldots, f_n(\tau)$ equals the $(n-1)$-fold symbolic product and its LT is the product of the LT's of the components:

$$f_1(\tau) * f_2(\tau) * \cdots * f_n(\tau)$$

$$= \int_0^{\tau} f_1(u_1)\, du_1 \int_0^{\tau-u_1} f_2(u_2)\, du_2 \int_0^{\tau-u_1-u_2}$$

$$\cdots \int_0^{\tau-u_1-\cdots-u_{n-2}} f_{n-1}(u_{n-1})\, f_n(\tau - u_1 - \cdots - u_{n-1})\, du_{n-1}$$

$$\risingdotseq \mathfrak{f}_1(z)\, \mathfrak{f}_2(z) \ldots \mathfrak{f}_n(z). \qquad \text{(A 28)}$$

REFERENCES

The list of references given below includes cited matter only. Additional literature is listed in the books of Syski (1960) and Saaty (1961), which both contain nearly exhaustive bibliographies. More recent publications may be found in Prabhu (1965) and Cohen (1969). Especially for Chapter 9 (Simulation) the bibliography in Naylor, T. H., Balintfy, J. L., Burdick, D. S. and Kong Chu (1966), *Computer Simulation Techniques*, John Wiley, New York, should be mentioned.

BAILEY, N. T. J. (1954), On queuing processes with bulk service, *J. Roy. Statist. Soc.*, Ser. B, **16**, 80.

BARLOW, R. E., and PROSCHAN, F. (1965), *Mathematical Theory of Reliability*, John Wiley & Sons, New York.

BROADHURST, S. W., and HARMSTON, A. T. (1949), An electronic traffic analyser, *P.O. Elec. Engrs. J.* **42**, 181.

BROCKMEYER, E., HALSTRÖM, H. L., and JENSEN, A. (1948), *The Life and Works of A. K. Erlang* (all of Erlang's papers translated into English), The Copenhagen Telephone Co., Copenhagen.

BUXTON, J. N. (ed.) (1968), *Simulation Programming Languages*, North Holland Publ. Co., Amsterdam.

CARSLAW, H. S., and JAEGER, J. C. (1953), *Operational Methods in Applied Mathematics*, Oxford Univ. Press, London.

COBHAM, A. (1954, 1955), Priority assignment in waiting line problems, *Operations Research* **2**, 70; a correction, *ibid*. **3**, 547.

COCHRAN, W. G., and COX, G. M. (1957), *Experimental Designs*, 2nd ed., John Wiley & Sons, New York.

COHEN, J. W. (1969), *The Single Server Queue*, North Holland Publ. Co., Amsterdam.

CONWAY, R. W., MAXWELL, W. L., and MILLER, L. W. (1967), *Theory of Scheduling*, Addison–Wesley, Reading (Mass.).

COPSON, E. T. (1935), *Theory of Functions of a Complex Variable*, Oxford Univ. Press, London.

COX, D. R. (1955), The analysis of non-Markovian stochastic processes by the inclusion of supplementary variables, *Proc. Cambridge Phil. Soc.*, **51**, 433.

CROMMELIN, C. D. (1932), Delay probability formulae when the holding-times are constant, *P.O. Elec. Engrs. J.* **25**, 41.

DIETRICH, G., and WAGNER, H. (1963), Traffic simulation and its application in telephony, *El. Commun.* **38**, 524.

DOWNTON, F. (1955), Waiting times in bulk service queues, *J. Roy. Statist. Soc.*, Ser. B, **17**, 256.

EMSHOFF, J. R., and SISSON, R. L. (1970), *The Design and Use of Computer Simulation Models*, Macmillan, Basingstoke, England.

ENGSET, T. (1918), Die Wahrscheinlichkeitsrechnung zur Bestimmung der Wählerzahl in automatischen Fernsprechämtern, *Elektrotechn. Z.* **31**, 304.

ERLANG, A. K. (1918), Solution of some problems in the theory of probabilities of significance in automatic telephone exchanges, *P.O. Elec. Engrs. J.* **10**, 189.

ERLANG, A. K. (1925), Application du calcul des probabilités en téléphonie, *Ann. P.T.T.* **14**, 617.

FISHER, R. A., and YATES, F. (1963), *Statistical Tables for Biological, Agricultural and Medical Research*, Oliver & Boyd, London.

FRY, T. C. (1928), *Probability and its Engineering Uses*, D. Van Nostrand Co., Princeton, N.Y.

GERSCHGORIN, S. (1931), Über die Abgrenzung der Eigenwerte einer Matrix, *Izv. Akad. Nauk SSSR, Ser. fiz.-mat.* **6**, 749.

GILTAY, J. (1950), Static and transient statistics in telephone-traffic problems, *Appl. Sci. Research*, Ser. B, **1**, 413.

HAANTJES, J. (1938), *Wiskundige Opgaven* (in Dutch).

HADAMARD, J. (1903), *Leçons sur la Propagation des Ondes et des Equations de l'Hydrodynamique*, Paris 13.

HAMMERSLEY, J. M., and HANDSCOMB, D. C. (1967), *Monte Carlo Methods*, Methuen, London.

HILLIER, F. S., and LIEBERMAN, G. J. (1967), *Introduction to Operations Research*, Holden-Day.

JAISWAL, N. K. (1968), *Priority Queues*, Acad. Press, New York.

JANSSON, B(IRGER) (1966), *Random Number Generators*, Victor Pettersons, Stockholm.

JENSEN, A. (1950), *Moe's Principle*, The Copenhagen Telephone Co., Copenhagen.

KENDALL, M. G., and BABINGTON-SMITH, B. (1939), *Tables of Random Sampling Numbers*, Tracts for Computers, XXIV, Cambridge.

KENDALL, D. G. (1951), Some problems in the theory of queues, *J. Roy. Statist. Soc.*, Ser. B, **13**, 151.

KESTEN, H., and RUNNENBURG, J. TH. (1957), Priority in waiting-line problems, *Koninkl. Ned. Akad. Wetenschap., Proc.*, Ser. A, **60**, 312 and 325.

KHINTCHINE, A. (1932), Mathematisches über die Erwartung vor einem öffentlichen Schalter, *Mat. Sbornik* **39**, 73 (Russian; German summary).

KLEYNEN, J. P. C. (1971), *Variance Reduction Techniques in Simulation*, doct. diss. Tilburg (The Netherlands).

KOSTEN, L. (1937), Über Sperrungswahrscheinlichkeiten bei Staffelschaltungen, *Elek. Nachr. Techn.* **14**, 5; in French: Sur les problèmes de blocage dans les multiples graduées, *Ann. P.T.T.* **26**, 1002.

KOSTEN, L. (1942), *Over Blokkerings- en Wachttijdproblemen*, doct. diss. Delft (The Netherlands, in Dutch; French and German summaries).

KOSTEN, L. (1948), On the measurement of congestion quantities by means of fictitious traffic, *Het P.T.T. Bedrijf* (The Netherlands) **2**, 15.

KOSTEN, L. (1948a), On the validity of the Erlang and Engset loss formulae, *Het P.T.T. Bedrijf* (The Netherlands) **2**, 42.

KOSTEN, L. (1952), On the accuracy of measurements of probabilities of delay and expected times of delay in telecommunication systems, *Appl. Sci. Res.*, Ser. B, **2**, 108 and 401.

KOSTEN, L. (1967), The "custodian's problem", *NATO Conf. on Queueing Theory, Lisbon* (1965), The English Univ. Press, London.

KOSTEN, L. (1970), Simulation in traffic theory, *Sixth Int. Teletraffic Congr.*, Munich.

KOSTEN, L., MANNING, J. R., and GARWOOD, F. (1949), On the accuracy of measurements of probabilities of loss in telephone systems, *J. Roy. Statist. Soc.*, Ser. B, **11**, 54.

LE GALL, P. (1962), *Les Systèmes avec ou sans Attente et les Processus Stochastiques*, Dunod, Paris.

MANUCCI, F., and TONIETTI, A. (1969), Traffic simulation in a telephone network via satellite with preassigned and demand assigned circuits, *Int. Conf. on Digital Communication*, London.

MOLINA, E. C. (1927), Application of the theory of probability to telephone trunking problems, *Bell System Tech. J.* **6**, 461.

MORSE, P. M. (1958), *Queues, Inventories and Maintenance*, John Wiley & Sons, Inc., New York.

NAYLOR, T. H., BALINTFY, J. L., BURDICK, D. S., and KONG CHU (1966), *Computer Simulation Techniques*, John Wiley & Sons, New York.

OLSSON, K. M. (1967), Some different methods in using Markov chains in discrete time applicable to traffic simulations and certain accuracy problems in this context, *Fifth Int. Teletraffic Congr.*, New York.

PALM, C. (1938), Analysis of the Erlang traffic formulae for busy-signal arrangements, *Ericsson Tech.* **6**, 39.

PALM, C. (1943), Intensity fluctuations in telephone traffic, *Ericsson Tech.*, no. 44, 1.

PALM, C. (1946, 1957), Research on telephone traffic carried by full availability groups; in 1946 in Swedish; English in 1957 in *Tele*, no. 1.

PHIPPS, T. E. Jr. (1956), Machine repair as a priority waiting-line problem, *Operations Research* **4**, 76 (Comments by W. R. VAN VOORHIS, *ibid.* **4**, 86).

POLLACZEK, F. (1930), Über eine Aufgabe der Wahrscheinlichkeitstheorie, part I, *Math. Z.* **32**, 64; part II, *ibid.* **32**, 729.

POLLACZEK, F. (1934), Über das Warteproblem, *Math. Z.* **38**, 492.

POLLACZEK, F. (1946), La loi d'attente des appels téléphoniques, *Compt. rend.* **222**, 353.

POLLACZEK, F. (1959), Application de la théorie des probabilités posées par l'encombrement des réseaux téléphoniques, *Ann. télécommun.* **14**, 165.

POLLACZEK, F. (1961), *Théorie Analytique des Problèms Stochastiques Relatifs à un Group de Lignes Téléphoniques avec Dispositif d'Attente*, Mem. Sci. Math., Gauthier-Villars, Paris.

PRABHU, N. U. (1965), *Queues and Inventories*, John Wiley & Sons, New York.

RIORDAN, J. (1953), Delay curves for calls served at random, *Bell System Tech. J.* **32**, 100.

RIORDAN, J. (1956), (Appendix to Wilkinson, 1956).

RIORDAN, J. (1962), *Stochastic Service Systems*, John Wiley & Sons, New York.

SAATY, T. L. (1961), *Elements of Queueing Theory*, McGraw-Hill, New York.

SEVASTYANOV, B. A. (1957), An ergodic theorem for Markov processes and its application to telephone systems with refusals, *Teoriya Veroyatnostei* **2**, 106 (Russian, English summary).

SMIRNOV, V. I. (1964), *A Course of Higher Mathematics*, Pergamon Press, New York.

SYSKI, R. (1960), *Introduction to Congestion Theory in Telephone Systems*, Oliver & Boyd, London.

SYSKI, R. (1967), Pollaczek method in queuing theory, *NATO Conf. on Queueing Theory, Lisbon* (1965), The English Univ. Press, London.

TAKÁCS, L. (1955), Investigation of waiting time problems by reduction to Markov processes, *Acta Math. Acad. Sci. Hung.* **6**, 101.

TOCHER, K. D. (1963), *The Art of Simulation*, Van Nostrand Co., Princeton, New York.

VAULOT, A. E. (1946), Délais d'attente des appels téléphoniques, traités au hasard, *Compt. rend.* **222**, 268.

VAULOT, A. E. (1954), Délais d'attente des appels téléphoniques, dans l'ordre inverse de leur arrivée, *Compt. rend.* **238**, 1188.

WILKINSON, R. I. (1956), Theories for toll traffic engineering in the U.S.A. (with an appendix by J. Riordan), *Bell System Tech. J.* **35**, 421.

AUTHOR INDEX

SUBJECT INDEX

OTHER TITLES IN THE SERIES IN PURE AND APPLIED MATHEMATICS